T0073739

The Dimming of Starlight

The Dimming
of Starlight

The Philosophy of Space Exploration

GONZALO MUNÉVAR

Professor Emeritus
Lawrence Technological University

OXFORD
UNIVERSITY PRESS

OXFORD
UNIVERSITY PRESS

Oxford University Press is a department of the University of Oxford. It furthers
the University's objective of excellence in research, scholarship, and education
by publishing worldwide. Oxford is a registered trade mark of Oxford University
Press in the UK and certain other countries.

Published in the United States of America by Oxford University Press
198 Madison Avenue, New York, NY 10016, United States of America.

© Oxford University Press 2023

All rights reserved. No part of this publication may be reproduced, stored in
a retrieval system, or transmitted, in any form or by any means, without the
prior permission in writing of Oxford University Press, or as expressly permitted
by law, by license, or under terms agreed with the appropriate reproduction
rights organization. Inquiries concerning reproduction outside the scope of the
above should be sent to the Rights Department, Oxford University Press, at the
address above.

You must not circulate this work in any other form
and you must impose this same condition on any acquirer.

CIP data is on file at the Library of Congress

ISBN 978–0–19–768991–2

DOI: 10.1093/oso/9780197689912.001.0001

Printed by Sheridan Books, Inc., United States of America

To my beloved son, Ryan, who has been asking me questions about space since he was a very little boy

Contents

Foreword

Human space exploration has slowed down in recent decades. At the same time, uncrewed space exploration is taking place quite intensively. The discussion of long-term human space missions to the Moon and Mars also is resurfacing. Of course, the first and fundamental question is how to accomplish such missions safely and efficiently. But no less important is the question: why do we need these missions, both uncrewed and human? Reflection on the motives behind the first space race was limited by the political and military rivalry between the two superpowers. The U.S. victory and achieving superiority in space led to a significant reduction in NASA's budget in the years following the Apollo program. In a sense, the situation became clear: since we don't need space for political purposes, it's not really clear why we need it, other than to conduct "casual" scientific research, as much as the budget allows. What characterized the post-Apollo era of space exploration was the lack of a grand, inspiring, and unifying idea and motivation for space exploration.

Today we again face this question about the meaning and purpose of our presence in space. The political rivalry between spacefaring countries plays a role, but it is incomparably less intense and dominant than the one during the Cold War. Moreover, humanity today has knowledge it did not have during the previous space race about global problems, with climate change at the forefront. In a sense, in the eyes of many, the idea of space exploration requires special justification and is not something to be taken for granted. Often, the cautious optimism of space exploration enthusiasts is limited only to robotic missions, while crewed missions are treated as an unnecessary, expensive, and dangerous extravagance. There is no doubt that in the third decade of the twenty-first century, the idea of advanced and long-term space exploration, especially human missions, requires a wise and comprehensive justification.

Such wise and comprehensive justification is an extremely difficult art. Repeating general clichés today is no longer enough. At this art of philosophical, in-depth justification for space exploration Gonzalo Munévar has perfectly succeeded. Munévar in his excellent book *The Dimming of Starlight: The Philosophy of Space Exploration* invites the reader on an exciting journey in which, step by step, the author reveals to the reader the crucial importance of space exploration for humanity. Munévar convincingly justifies why we must consider space as the next stage of our development as a species. As not only a philosopher of space exploration but also a philosopher of science, Munévar devotes much attention to justifying space exploration from the point of view of its importance to science. Significantly, the author is one of the few to show that science in and about space is not just an applied science, as is usually presented, but a fundamental science. This is, in my opinion, the most important value of this book and its contribution to the philosophy of space exploration. Munévar has succeeded in showing that space exploration has a fundamental, crucial value for humanity, including scientific understanding of ourselves, the world, and the universe.

The author also points out numerous applicable aspects of scientific space exploration. Among the author's many interesting and convincing arguments, it is worth mentioning one where the author shows why space exploration is also important for solving global problems. This is a point that should be clarified by any space exploration enthusiast confronted with the skeptical arguments of social critics raising the profile of pressing problems on Earth, against which advanced space exploration seems an unjustified waste of resources. Munévar points out that the space perspective treats the Earth as one global system, so that we acquire the right perspective to save the Earth, which is, after all, a global problem and task. Secondly, the author emphasizes that by learning about other worlds, we can better know and understand (and appreciate) our own world.

The Dimming of Starlight is a much-needed book among monographs on the philosophy of space exploration. It is today that space exploration requires a special comprehensive justification, especially in light of global problems on Earth. It is worth noting that Munévar has not only provided such a justification, which should convince even the staunchest opponents of space exploration. Perhaps more importantly, the author has presented universal arguments in favor of space exploration. Opponents of space exploration should feel persuaded to change their skeptical position and see the possible errors in their reasoning that Munévar enumerates. Space enthusiasts, on the other hand, receive into their hands an effective weapon and several new arguments in justifying their position in favor of the strategic importance of space exploration for our species.

Konrad Szocik

Preface

This book is the result of an intellectual adventure into the nature and justification of space exploration. That adventure had the most auspicious beginnings many years ago. It started with a 1979 grant from the National Science Foundation to develop and team-teach a course on space exploration with my colleague John C. Kasher at the University of Nebraska at Omaha. Jack and I stood together before our upper-division and graduate students, taking turns—he the physicist, I the philosopher of science—to explain a point or to debate some controversy about the subject. The grant also paid for the visit of a prominent figure in the field; and we were fortunate to bring the Princeton physicist Gerard O'Neill, author of the very famous book *The High Frontier*, who gave a most extraordinary presentation. He also gave Jack and me the opportunity for more private discussion. At a lunch, I took advantage of O'Neill's agreeable disposition to bring my one-year-old son, Ryan, to be in the presence of someone who might become the Columbus of outer space. Little we knew of the semi-paralysis the Space Shuttle would bring to our space adventure. After our very successful course, NASA published our account of it. And every other year after that, up to 1989, Jack and I taught the course again, to our satisfaction and that of our students.

The next development was even more auspicious. In the academic year of 1983/1984, I became a Fellow at the Stanford Humanities Center, with a project centered on the philosophy of space exploration. Before going to Stanford, I spent the summer at the University of Maryland in a National Endowment for the Humanities seminar on the history of science, directed by Stephen G. Brush, a famous historian of physics and astronomy with a great interest in space science (see Chapter 5). Once at Stanford, the treasures of space exploration spread all around me. I was enlightened and mesmerized by my many discussions with space scientists from NASA Ames Research Center, just a few miles from my residence. Indeed, Christopher McKay, who has provided fascinating proposals for terraforming Mars, was one of the first to learn that my book would be titled *The Dimming of Starlight*. I also remember talking to Brian Toon, Steve Squyres, and Harold Kline (all mentioned later in the book). Several researchers in the biology section were also very welcoming and helpful, and rather excited about my first chapter, as I recall.

I have a cherished memory of Harold Kline, who was the director of biology of the Viking experiments to find life on Mars. In the winter of 1984, I was teaching my graduate seminar at Stanford and Harold was scheduled as a guest speaker. But then Carl Sagan announced a visit to Ames, and Harold had to be his host. On that day, I went to hear the first part of Sagan's talk, which was introduced by Harold. During the break, however, Harold turned the meeting over to someone else, and followed me on the freeway to the physics building at Stanford, where we arrived just in time for his presentation at my seminar.

That seminar, which gave exposure to the first rough draft of my book, was one of the great highlights of my academic career. It was attended by several Stanford scientists who were very supportive of me during that year. I very fondly remember the participation of Ronald N. Bracewell, an extraordinary space scientist who had written on how intelligent aliens could observe us unobtrusively. And then, if I remember correctly, by the Associate Dean of Engineering, James F. Gibbons, who had taken me under his wing earlier, and had included me in a group that held discussions on science and technology. At one point the guest speaker was none other than Freeman Dyson! At any rate, at the seminar meetings there were often more faculty members than graduate students, joining us for a spirited discussion. It was not unusual to have two or three of the top experts in the world on the topic for the day, between Stanford and Ames luminaries.

Other Stanford faculty took it upon themselves to be helpful as well. Pierre Noyes would invite me to lunch at the Faculty Club to make sure, over the excellent meals, that I really understood rockets (see Chapter 9) and to intrigue me with his ideas about the subatomic world. Francis Everitt took me to his lab to see his most recent model of the perfect sphere that would so rigorously test Einstein's general theory of relativity in space (see Gravity Probe B, Chapter 5). Even Gerard O'Neill was there. Although he was a professor at Princeton, his experimental life was centered on the Stanford Linear Accelerator, for he had invented a way of having beams of particles collide with each other head-on by directing them on opposite directions in a magnetic ring at the end of the long accelerator (the RINK). The collisions would be far more energetic than hitting a static target, thus producing abundant new particles to study. Many thought it was an accomplishment worth the Nobel Prize. At any rate, he was again generous with his time, and even treated me to a tour of the Stanford Linear Accelerator.

During that extraordinary year, I gathered much evidence for an important insight my book offers: that exploring space will transform our views of the Earth and the universe in fundamental ways, and to the significant benefit of our species. As we explore space, we challenge our science, and as we challenge our science, we might change it in ways so profound that we come to face a different panorama of problems and opportunities in our dealings with the universe. Indeed, it is as if a new universe opens up to us; and when we try to adapt to the new "lay of the land," ideas and inventions occur to us that would have been unimaginable under the old perspective.

It became clear to me then that serendipity is a natural, practically inevitable consequence of scientific exploration. My argument thus depends on the very nature of scientific exploration and on the way that nature is illustrated in space science and other aspects of our space adventures. In Chapters 4, 5, and 6, I point out that such serendipity is to be found in the contribution of planetary exploration to our knowledge of the Earth, on the fundamental changes to physics that space studies of cosmology have created, as well as on the scientific benefits that derive from astrobiology even if we never find any extraterrestrial life. And there is much more.

After such an auspicious beginning, perhaps the many papers and conference presentations, based on or at least inspired by my 1984 draft, should have soon been followed by the publication of the book itself, but unfortunately many nonacademic incidents created major impediments. The worst was the year I devoted to what was supposed to be my final draft. The computer I had bought for the purpose suddenly kept me from copying the files into diskettes. In those days that brand had very poor support. I was told to mail the computer to some center, to be returned who knew when. I could not lose that time, so I kept on. And one morning, when I was very near the end, I turned on the computer only to be met with a large drawing of a frown. I had lost every file! I thought I would rebuild from previous drafts, now part-time. But the relevant diskettes were themselves damaged in the summer heat that bathed the truck I used in a move. After that, health problems that made me unable to type seemed to show up just every time I managed to get some released time for working on my book. I came to suspect that *The Dimming of Starlight* was jinxed. But I have finally come through, even if it took me decades. I do hope that this very long struggle will be shown to have been worthwhile by the pages you are about to read. I should clarify, though, that the present version has evolved considerably compared to what I would have published decades ago. Had I published then, this would be the second, updated edition.

Acknowledgments

I wish to acknowledge the contributions of the several people who have helped bring about the final draft of this book. I thus thank Dr. Jill Banfield of U.C. Berkeley for the great pictures of ultra-small bacteria, my former student Ruoyu Huang for her terrific illustration, and Dr. Michael R. Shapiro for his four beautiful astrophotographs. This book would not have been possible without the computer help of my colleague Dr. Matthew Cole, who provided clean final drafts of all the chapters as well as a PDF of the whole book. My former student Phillip McMurray has been extraordinarily helpful, solving computer problems and formatting text and images, initially, and then in preparing the manuscript for publication. He has also provided many useful questions and comments, as well as contributing the Index. On the matter of comments, I am extremely grateful to Drs. David Paulsen and Konrad Szocik for their very careful, and I am sure time-consuming, reading of my manuscript, and for their advice on how to improve it, and to Dr. Szocik for his very flattering Foreword to the book. I benefitted as well from years of discussion with my M.A. advisor at Northridge, Dr. Daniel Sedey. I also wish to thank my wife, Dr. Susan Greenshields, for her encouragement and her talent for finding typos. I reserve a special gratitude for all the students, graduates and undergraduates, who contributed their questions and enthusiasm for my space exploration courses at the University of Nebraska at Omaha, Stanford University, the University of California at Irvine, Evergreen, and Lawrence Technological University. And, finally, I should thank all the people who have been writing to me over the years expressing their hope for the culmination of my efforts on the subject.

My drafts of *The Dimming of Starlight* have led to the publication of several papers that in turn allowed me to sharpen my views. I acknowledge here the relevant publishers for thus facilitating the improvement of my book. "Philosophy, Space Science and the Justification of Space Exploration," *Essays on Creativity and Science*, ed. Diana M. DeLuca, HCTE (Hawaii, 1986), pp. 89–96. "Pecking Orders and the Rhetoric of Science," *Explorations in Knowledge* III, no. 2 (Spring 1986): 43–48. "Rhetorical Grounds for Determining What Is Fundamental Science: The Case of Space Exploration," in *Argument and Social Practice*, ed. J. R. Cox, M. O. Sillars, and G. W. Walker (Speech Communication Association, 1985), pp. 420–434. "Space Colonies and the Philosophy of Space Exploration," in *Space Colonization: Technology and the Liberal Arts*, ed. C. H. Holbrow, A. M. Russell, and G. F. Sutton, *American Institute of Physics, Conference Proceedings 148* (1986): 2–12. "Filosofía y la Evaluación de la Tecnología Espacial," *Arbor* 509, no. CXXX (May 1988): 59–72. "Human and Extraterrestrial Science," *Explorations in Knowledge* 6, no. 2 (1989): 1–9. "A Philosopher Looks at Space Exploration," *Evolution and the Naked Truth* (Ashgate, 1998), pp. 169–179. "Philosophy and Exploration of the Solar System," *Philosophic Exchange*, no. 28 (1997–1998): 56–61. "Venus y el Fin del Mundo," *Eidos*, no. 4 (March 2006): 10–25. "Humankind in Outer Space," *The International Journal of Technology, Knowledge and Society* 4, no. 5 (2008): 17–25. "Einstein y el límite de la velocidad de la luz," in *Einstein: Científico y filósofo*, ed. G. Guerrero (Programa Editorial Universidad del Valle, 2011), pp. 291–308. "Self-Reproducing Automata and the Impossibility of SETI," in *Imagining Outer Space: European Astroculture in the Twentieth Century*, ed. A. C. T. Geppert (Palgrave MacMillan, 2012), pp. 267–284. "Space Exploration and Human Survival," *Space Policy* 30 (Nov. 15, 2014): 197–201. "Space Colonies and Their Critics," *The Ethics of Space Exploration*, ed. James S. J. Schwartz and Tony Milligan (Springer, 2016), pp. 31–46. "An Obligation to Colonize Outer Space," in *Human Colonization of Other Worlds. Futures*, ed. Kelly C. Smith and Keith Abney, 110 (June 2019), pp. 38–40. Kelly C. Smith, Keith Abney, Gregory Anderson, Linda Billings, Carl L. DeVito, Brian Patrick Green, Alan R. Johnson, Lori Marino, Gonzalo Munévar, Michael P. Oman-Reagan, Adam Potthast, James S. J. Schwartz, Koji Tachibana, John W. Traphagan, and Sheri Wells-Jensen, "The Great Colonization Debate," *Futures* 110 (June 2019): 4–14. "Science and Ethics in the Exploration of Mars," in *The Human Factor in a Mission to Mars: An Interdisciplinary Approach*, ed. Konrad Szocik (Springer, 2019), pp. 185–200. "Ethical Obligations towards Extraterrestrial Life," *Philosophy Study* 10, no. 3 (March 2020): 193–201, doi:10.17265/2159-5313/2020.03.003. "Science and Ethics in the Human-Enhanced Exploration of Mars," in *Human Enhancements for Space Missions: Lunar, Martian, and Future Missions to the Outer Planets*, ed. Konrad Szocik (Springer, 2020), pp. 113–124. Konrad Szocik, Mark Shelhamer, Martin

Braddock, Francis A. Cucinotta, Chris Impey, Pete Worden, Ted Peters, Milan M. Ćirković, Kelly C. Smith, Koji Tachibana, Michael J. Reiss, Ziba Norman, Arvin M. Gouw, and Gonzalo Munévar, "Future Space Missions and Human Enhancement: Medical and Ethical Challenges," *Futures* 133 (2021): 185–200.

The Dimming of Starlight

1

Why Philosophy?

One night almost four hundred years ago, Galileo turned his telescope to the sky, and the sky grew immense and crowded (see Figure 1.1). Since then, we have explored the heavens with telescope and mind, in the spirit of wonder and adventure. In our own time, through space exploration, we can touch where Galileo could only see, and we can reach where he could only dream. Our spaceships are beginning to realize a perennial longing made explicit by the great astronomer Johannes Kepler when he wrote to Galileo:

> There will certainly be no lack of human pioneers when we have mastered the art of flight. . . . Let us create vessels and sails adjusted to the heavenly ether, and there will be plenty of people unafraid of the empty wastes. In the meantime, we shall prepare, for the brave sky-travelers, maps of the celestial bodies—I shall do it for the Moon, you, Galileo, for Jupiter.[1]

Sky-travelers are, at long last, sailing along the routes marked on the maps of Kepler and Galileo. And as Kepler would have imagined, they find adventure, beauty, and excitement in the enterprise. They also promise us knowledge and bright new hope if humankind agrees to expand first into the solar system and eventually into the galaxy. But how firm is this promise? And what sacrifices should we make so that it can be kept? Those are the main questions of this book. I want to examine why human beings explore space and to determine whether we ought to.

This examination is by no means easy, for space exploration elicits many polemical responses. On the one hand, we have the enthusiasm of people like Wernher von Braun, the famous rocket expert, who claimed that "[T]he first moon landing was equal in importance to that moment in evolution when aquatic life came crawling on the land."[2] On the other hand, we have social and ideological critics. The social critics argue that we are besieged by illness, poverty, and hopelessness. We thus have an obligation to invest our money, talents, and resources to solve these human problems, but the pursuit

The Dimming of Starlight. Gonzalo Munévar, Oxford University Press. © Oxford University Press 2023.
DOI: 10.1093/oso/9780197689912.003.0001

Figure 1.1. The Pleiades Star Cluster (M45). Located about 444 light years away, M45 is one of the closest star clusters to Earth. Although easily visible to the naked eye, Galileo was the first to observe the cluster through a telescope and found many more stars too dim to see without visual aid. The blue color is light reflected off the surrounding dim dust from the hot blue stars of the cluster. This image is about ten hours of data captured with a Celestron RASA 8 (see color plate). (Image courtesy of Michael R. Shapiro)

of space exploration competes for the means needed to fulfill our obligation. The ideological critics view space exploration as a logical extension of science, and science (at least "big science") as a basically unwise activity, for science leads us to interfere with nature instead of trying to live in harmony with it. According to them, this now massive interference has brought the world to the brink of environmental catastrophe. Only a change of ideology, or perhaps of moral outlook, can give us hope. The "promise" of space is then nothing but a siren song that diverts our attention at a crucial moment in our history.

In response to these and other critics, space enthusiasts list the many benefits we derive from the space program: weather satellites save lives and crops, communication satellites bring about economic expansion, and land satellites discover resources and help us monitor the environment. Moreover, space technology spins off valuable products into our lives, such as reflective insulation and voice-controlled wheelchairs.

Why Explore Space? An Overview

Why then is space exploration adrift? And why does it no longer excite the public passion as it did during its Golden Age in the 1960s, when we went to the Moon and the sky was no longer the limit? Should not the response by the space enthusiasts light star fires in the eyes of their fellow citizens? Why do the enthusiasts' arguments fail to align social policy with their values and dreams? Econometric studies have not done the job. Comparisons of (presumed) costs and benefits have not done the job. Why does the bulk of humankind remain blind to such wonderful treasures at the end of cosmic rainbows? At the time of this writing, it seems that space exploration is enjoying a renaissance, both in enthusiasm and activity, and it is therefore crucial that these questions be addressed satisfactorily, so that such a renaissance will indeed leave a deep mark in history.

Part of the reason for the ending of the first Golden Age has to do with the bad choices made since the National Aeronautics and Space Administration (NASA) became one more sluggish bureaucracy, particularly since its fateful decision to build the Space Shuttle, as I argue in Chapter 7. But the main reason is that space enthusiasts have not offered enough of a compelling argument. As we will see in Chapter 2, the social critics may simply accept space exploration but only to a point, as in fact most people do. They will agree, for example, to the likes of communication satellites, from which we clearly derive benefits. But the heart of space exploration is found beyond such obviously practical pursuits: daring space missions such as the probes of Jupiter and Titan give us knowledge, and, yes, that knowledge is exciting, but is it better than improving the lives of people? Thus, we have the same objection again, even if the scope is somewhat reduced. As for the ideological critics, they will stick to their guns, continuing to argue that the problems that our adventures in space might help alleviate would not arise if we learned to treat our environment and each other differently.

Space enthusiasts like to appeal to the unintended benefits of previous scientific exploration. Who could have imagined so many serendipitous discoveries when the first human-made satellite, Sputnik I, went into orbit in 1957? But can we really trust the promise that our most esoteric and daring adventures will deliver new and presently unimagined bounty? As we will also see in Chapter 2, the historical anecdotes generally offered to support the notion of the serendipity of science are not enough.

Can we offer enough? Yes—enough indeed to justify the exploration of space, as I argue in Chapter 3. We may begin by noticing that each side of the controversy justifies its position by appeal to the things it values, and that each stresses different values. The issue of justification thus has the air of a philosophical problem. And so it is, though not because it is a hopeless muddle, but because philosophical tools can be deployed to resolve it. Of these tools, the first is the philosopher's search for the assumptions that underlie the problem. Eventually this search will lead us to the realization that they are assumptions about the nature of science.

For example, the social critics find the value of scientific knowledge—as obtained through space science—not large enough to justify the money that it presumably takes away from attending to other human needs. But to estimate the value of scientific knowledge in any fruitful way, one should have some idea of what science is like and of what it has to offer.

The ideological critics, for their part, hold that science is unwise. But what insights about science have led them to such a conclusion? And since reflecting on the nature of science is the province of the philosophy of science—whether done by philosophers, scientists, or laypeople—the resolution of this important controversy in scientific and social policy is also a job for the philosophy of science.

My own reflections lead me to conclude that we ought to explore space. One crucial reason, as I argue in Chapter 3, is that the exploration of space will transform our views of the Earth and the universe to the significant benefit of our species. As we explore space, we challenge our science, and as we challenge our science, we change it in ways so profound that we come to face a different panorama of problems and opportunities in our dealings with the world. Indeed, it is as if a new world opened up to us; and when we try to adapt to the new "lay of the land," ideas and inventions occur to us that would have been unimaginable under the old perspective.

We will see, in other words, that serendipity is a natural, practically inevitable consequence of scientific exploration. My argument will thus depend on the very nature of scientific exploration and on the way that nature is illustrated in space science and other aspects of our space adventures.

Space Exploration as Fundamental Science

The notion that science and space exploration go hand in hand may seem obvious to a casual observer, but it has been bitterly contested over the years. Many scientists, perhaps the majority of scientists, were opposed to the Apollo program, to put a man on the Moon, on the grounds that it was political showbiz and not science. And just about every important field of space science has been denigrated, at one time or another, in the most prestigious and established quarters of science. Some of those fields still are.[3] And if we pay attention, we may still hear rumblings that all that money should go for truly important research. Indeed, a common complaint until recently, particularly in the physical sciences, has been that space science is merely applied science, and thus it would follow that, if we wish to forge changes to our fundamental views of the world, we should concentrate on putting our money and effort into fundamental science, not into space science.

In my reply I will show how every main branch of space science leads to new perspectives of immense value. I will argue in Chapter 4 that several of the main problems that our planet confronts now (e.g., the depletion of the ozone layer and global warming), as well as those it will probably confront in the next few centuries, are far more likely to be solved thanks to space exploration in two ways. The first is that such problems tend to be global problems and space technology is particularly well suited to study the Earth as a global system. The second is that as we explore other worlds, we gain a broader and deeper understanding of our own planet.

From comparative planetology we will move on to space physics and astronomy, two fields ripe with the promise of radical changes to our scientific points of view. Such changes will in turn yield an extraordinary new harvest of serendipitous consequences for technology and for our way of life. The reason these two fields are ripe with promise is simple. The Earth's atmosphere limits drastically the information we receive about the universe because it blocks much of the radiation that comes in our direction. This shielding is, of course, a good thing, for otherwise life could not exist on our planet. But to make even reasonable guesses about the nature of the universe, we need that information. That is why we need telescopes in orbit and eventually on the Moon and other sectors of the solar system, as the James Webb space telescope demonstrates. Until the day when space telescopes began to operate, many physicists thought of space physical science as applied science, mere application, that is, of the very successful "standard model" that explained matter in term of its constituting particles and the forces between them.

But, as I discuss in Chapter 5, physicists had been trying to explain a limited universe—a universe based on what we could observe through a few peepholes in the walls that protected us from cosmic dangers. It had already been known for some time, though not widely, that the visible mass in galaxies did not exert enough gravitational force to keep their outer rims of stars from being flung into intergalactic space. Astronomers presumed that eventually the missing mass would be found, but when space telescopes gave us the whole electromagnetic spectrum to look for that mass, we still could not find enough of it. According to some high estimates, up to 90% of the mass needed to account for the behavior of galaxies is undetectable ("dark matter"), and radically different from the matter explained by the "standard model."

To make a bad situation worse, in the late 1990s space astronomers discovered that the expansion of the universe was accelerating, even though we should expect that, after the Big Bang, gravity would slow down the rate of expansion. A new form of energy ("dark energy") is supposed to explain this bewildering state of affairs, once we determine what its properties are.

Fundamental physics, which uses the "standard model" to think about the universe, explains *familiar* matter and energy. But most of the universe seems to be made up of *unfamiliar* dark matter and energy, perhaps even upward of 95% when you combine those two. This means that thanks to space science, we found out the extraordinary extent of our ignorance, and that space science is a necessary tool for developing a new physics.

Moreover, as space science opened new kinds of phenomena to our attention, astronomers began to develop new observational technologies to compensate for the atmosphere, so as to explore the universe better from the surface of our planet.

Space exploration is also ripe with promise for biology, as we will see in Chapter 6. This promise is particularly interesting in the case of the astrobiologists' attempt to search for life in other worlds. For example, when a NASA team announced in 1996 that a Martian meteorite contained organic carbon and structures that looked like fossils of bacteria, meteorite experts adduced that inorganic processes could account for all the substances and structures found in the meteorite. Therefore, these experts claimed, by Occam's razor, we should reject the (ancient) Martian-life hypothesis (Occam's razor is a principle that favors the simpler hypothesis; it is named after William of Occam, a medieval philosopher). Other scientists pointed out, in addition, that the presumed fossils were about one hundred times smaller than any known bacteria, too small in fact to be able to function as living organisms. But as we will see in Chapter 6, Occam's razor would, if anything, favor the Martian-life hypothesis; *and,* ironically enough, the claim about the minimum size of living things spurred a rather controversial search that led, first, to the discovery of extremely small structures that reproduce and mimic life in some other respects, and then of ultra-small bacteria.

This is one example of how investigating the possibility of extraterrestrial life leads to the improvement of the science of terrestrial biology. I will also provide a more general argument.

Space biology proper (doing biological experiments in space) has not yet produced spectacular and significant discoveries, but, as we will also see in Chapter 6, the main objections against its scientific value are based on misguided distinctions between fundamental and applied science not unlike those advanced some years ago against the space physical sciences. Some of these objections are also based on mistaken assumptions about genetics, and particularly about the relationship between genotypes and phenotypes.

Space scientists, who may be generally sympathetic to the main theses of this book, are nevertheless deeply divided on the question of how best to explore space. Some claim that exploring with humans is frightfully expensive and dangerous, that the Space Shuttle has set back the cause of exploration, and that continuing to favor astronauts over robot spacecraft will set it back even further. And they are indeed correct—in the short run. I argue in Chapter 7 that a measured increment of the human presence in space will eventually lead to even greater opportunities for all the space sciences. I also point out how the proposed colonization of other planets, the mining of the asteroids, and the expansion into the outer solar system, and perhaps the galaxy, may secure the survival of the human species. Of course, such fanciful proposals may be little more than far-fetched dreams, but those dreams begin to pull us away from our mother planet, and as they color our perception of space exploration, they influence its direction. Even more fanciful, although of special scientific and philosophical interest, are the heated debates about relativistic starships and faster-than-light travel.

Perhaps no aspect of space exploration has been as controversial as the search for extraterrestrial intelligence (SETI). For some it has been a noble calling; for others it has been the most ridiculous waste of money and effort. The critics won the day in Congress when NASA was forced to drop SETI altogether many years ago, although private donations and platoons of volunteers have kept the search going. As we will see in Chapter 8, many of the arguments for and against the existence of extraterrestrial intelligence are based on what Carl Sagan called the "Principle of Mediocrity" (that the Copernican revolution has taught that there is nothing special about the Earth or its place in the universe). But, as I will argue, such a principle does not stand up to criticism. We have no good reasons for optimism or pessimism on this matter: the most reasonable position is agnosticism.

This is not to say that SETI is a worthless enterprise. For example, the problem of how we might communicate with extraterrestrial civilizations, if there are any, teaches us a few things about how we understand the world and ourselves. It is often thought that advanced species will have discovered many of the fundamental laws of physics, chemistry, and so on; otherwise, they could not make the attempt to communicate across the vastness of interstellar space. But since the laws of nature are (presumably) the same everywhere, and since they are expressed in mathematics, all advanced species will have things in common that can serve as the basis of communication. According to this conventional wisdom, then, there must be intellectual convergence between highly intelligent species, just as there is convergence of form between fishes and dolphins.

But how can we support this assumption of convergence? Evolutionary history is made up of millions of contingencies. It would be practically impossible for life to evolve in other worlds along the same paths it has followed on Earth. We thus face an unpleasant consequence: a different evolutionary history *may* produce different brains—different ways, that is, of perceiving the environment and of putting those perceptions together. And those are the brains that will one day develop science. It is thus plausible to suggest that those brains will operate with mental categories different from ours, and that alien science and mathematics may also differ from ours. Discussing the assumption of convergence will thus involve us in the philosophical problem of whether we discover or invent science.

Another idea whose discussion leads to a better understanding of living beings is the suggestion by Freeman Dyson and others that we should use von Neumann's self-reproducing automata to colonize the galaxy. I argue, also in Chapter 8, that the very idea of such technology is based on the mistaken metaphor of the genome as a computer program. The speculations by Robert Zubrin that nanotechnology will allow us to get around the overwhelming obstacles to self-reproducing automata do not get very far either, for some of the most fanciful claims made about nanotechnology are also without justification.[4]

Many interesting issues come up in the details of practically all the fields of exploration discussed in this book. In Chapter 4, for example, I note that an argument against the possibility that Venus once had oceans has the same structure as an argument for the end of the world (or more precisely, of humankind) advanced by the philosopher John Leslie and inspired by the physicist Brandon Carter's account of the anthropic principle. In my opinion, both the objection to Venusian oceans and Leslie's argument assume an untenable view of probability.

Whatever the benefits of space exploration, it also involves a variety of risks. One danger, in particular, seems to be of great importance: the unavoidable connection between space technology and war. This connection is presumably made quite obvious by the terror inflicted upon London in World War II by Wernher von Braun's V2 rockets and strengthened by Ronald Reagan's proposal for a "Star Wars" defense against the Soviets' intercontinental ballistic missiles, themselves strong evidence of the evils humans fall prey to when reaching for the heavens. We will see in Chapter 9, however, that the connection between space technology and war is not quite that obvious. Its apparent plausibility comes from popular historical interpretations of the relevant episodes, but a closer look fails to support the claim that the connection is unavoidable. Moreover, space technology may prove to be key to the long-term survival of terrestrial life, as Zubrin and others have claimed.

By Chapter 10, it will be clear that the profound practicality of science, via the serendipity that is its natural consequence, provides an adequate response to the social critics. Our new understanding of science, in light of space exploration, will also set aside the concerns of the ideological critics. Most ideological criticisms stem from purported insights about the relationship between human beings and the environment of the Earth—insights such as the balance of nature, the wisdom of noninterference with natural processes, and so on. But as we will see, such insights do not withstand scrutiny. Moreover, to offer a strong argument, the ideological critics need a global understanding of the Earth's environment. But as I explain again in this final chapter, that global understanding requires the assistance of comparative planetology and space technology. To meet their ultimate goals, and our obligation to future generations, they would do well to ally themselves with the "big science" they so often deride.

Space Exploration as a Subject for Philosophy

It is surprising that professional philosophy has not, until recently, taken up in earnest the question of whether space exploration can be justified. After all, it is part and parcel of philosophy to examine the justification of all significant human activities, including science and technology. Professional philosophers have examined, for example, what computers can and cannot do,[5] whether human cloning should be permitted,[6] and the extent to which even theoretical scientists are morally responsible for their research.[7] But the call of the "final frontier," whose significance may well dwarf all other human adventures, has received relatively scant attention from philosophy. The lack of interest is due in part to the fact that research on foundations has the highest prestige in philosophy, and a field named "the philosophy of space exploration" probably sounds like applied philosophy to most practitioners. But the distinction between fundamental (or pure) and applied can be as misleading in philosophy as in science. This can be seen, for example, in the case of the philosophy of artificial intelligence, where it became clear that the mind is not at all like a digital computer—a fundamental finding in philosophy.[8] I would argue that the deep practicality of science, found in the examination of space science, is also a fundamental finding. Whatever the reason professional philosophy has not concerned itself with the exploration of space, this book is an attempt to remedy that oversight. In this, I hope it will complement the growing literature on the ethics, as well as on other philosophical and social aspects of space exploration.[9]

Scientific textbook mythology would have us believe that Galileo's opponents simply refused to believe their own eyes when looking through his telescope (or worse, refused to even look through it!). But the documents of that time indicate instead that many who looked through Galileo's telescope saw double images or the disc of the Moon displaced to one side, and some people could see nothing at all where Galileo claimed the existence of Jupiter's moons. This is not surprising. When we perceive, we make use of many clues from the environment. We estimate at a glance, say, how large a new painting is, because we see that it takes up so much of a familiar wall. Without those clues, our vision falls prey to all sorts of illusions. For example, a point of light on a dark background may appear to move around rapidly. That is why Jupiter shining through a fog in the early evening is often reported as a UFO. Consider also that Galileo's telescope was primitive and built for his eyes (it lacked a focus mechanism). And to make matters worse, the testimony of the telescope often conflicted with that of the unaided eye.[10]

One of those conflicts involved the brightness of starlight, for in Galileo's telescope the magnitude of the stars did not keep pace with that of the planets (planetary disks were enlarged, whereas stars remained points of light). Some observers even complained that his telescope made the stars look dimmer.[11] The dimming of starlight was then one of the illusions that threatened the new scientists' exploration of the cosmos. Four hundred years later, as we seek to launch human and machine alike to new worlds, there are those who recoil at the very thought. Where vision once undercut the exciting appeal of the new, today social and ideological misgivings would keep us from reaching for the heavens.

Supporters of space exploration sometimes find favorable omens in the insights of long ago. They should be delighted by Ovid's account of the difference between man and beast: "God elevated man's face and ordered him to contemplate the stars." It is for us now to determine whether Kepler was right in thinking that we should go beyond contemplation and reach for the stars. If he was, we should not let their light dim again.

Notes

1. Johannes Kepler, *Conversations with the Star Messenger*, 1610. Partially quoted in A. Koestler, *The Watershed: A Biography of Johannes Kepler* (University Press of America, 1960), p. 195.
2. Quoted in W. S. Bainbridge, *The Spaceflight Revolution* (John Wiley & Sons, 1976), p. 1.
3. These points will be discussed in detail in Chapters 3–7. Stephen G. Brush has aptly illustrated the significance of the space sciences to the development of physics, as will be seen particularly in Chapter 5.
4. Robert Zubrin's seminal ideas about exploration will be discussed in several other chapters, particularly in Chapter 7.
5. The first classic work in this field was H. Dreyfus, *What Computers Can't Do: A Critique of Artificial Intelligence* (Harper, 1972). For a work on the nature of mind that includes neural nets (and thus parallel as opposed to serial processing), see P.

Churchland, *The Engine of Reason, The Seat of the Soul* (MIT Press, 1975). Part of the discussion in Chapter 8 will be based on insights derived from these works.

6. See, for example, Inmaculada de Melo-Martín, *Taking Biology Seriously: What Biology Can and Cannot Tell Us about Moral and Public Policy Issues* (Rowman and Littlefield, forthcoming).

7. See G. Munévar, "The Moral Autonomy of Science and the Recombinant DNA Controversy," *Journal of Social and Biological Structures* 2 (1979): 235–243.

8. Dreyfus, *What Computers Can't Do*; Churchland, *The Engine of Reason*.

9. See, for example, the collection of papers in *Space Policy* 30 (Nov. 15, 2014); James S. J. Schwartz and Tony Milligan (eds.), *The Ethics of Space Exploration* (Springer, 2016); A. C. T. Geppert (ed.), *Imagining Outer Space: European Astroculture in the Twentieth Century* (Palgrave MacMillan, 2012); Kelly C. Smith and Keith Abney (eds.), "Human Colonization of Other Worlds," *Futures* 110 (June 2019); Konrad Szocik (ed.), *Human Enhancements for Space Missions. Lunar, Martian, and Future Missions to the Outer Planets* (Springer, 2020); and James S. J. Schwartz, *The Value of Science in Space Exploration* (Oxford University Press, 2020).

10. P. K. Feyerabend, *Against Method*, 3rd ed. (Verso, 1993), Ch. 9.

11. As described in Feyerabend's *Against Method*, pp. 92–93, n26.

2

The Standard Case For and Against Space Exploration

Supporters of space exploration are often perplexed by the failure of their fellow citizens to grasp the significance and fascination of exploring the heavens. In the words of Arthur C. Clarke: "The urge to explore, to discover, to follow knowledge like a sinking star" is its own justification.[1] This self-justification is presumably rooted in human nature or human destiny. As the Norwegian explorer and Nobel laureate Fridtjof Nansen once said, "The history of the human race is a continuous struggle from darkness to light. It is therefore of no purpose to discuss the use of knowledge—man wants to know and when he ceases to do so he is no longer man."[2]

For supporters of exploration such as these, the matter is quite simple. Since knowledge is of the essence for humans, humans cannot help wanting to explore. Thus, it is destiny that propels us forward into the cosmos. Under such circumstances there is little point in bothering with a more extensive justification. We need only grab history by the tail and make it go where we wish to take it. Why should philosophy force us into reflection when action is so tempting?

One reason for reflection is that appeals to human nature and destiny are not convincing enough. In saying that it is in our nature to explore or that humanity's destiny is in the stars, we can mean several things, but some interpretations are more central than others. We can mean any of the following:

1. We are going to explore space come what may.
2. Because of our nature, we have a strong tendency to explore space.
3. Our nature is such that in some important sense we will not fulfill ourselves if we do not explore space.

Let us see how these interpretations fare in the task of making clear that space exploration is a worthwhile undertaking.

The Dimming of Starlight. Gonzalo Munévar, Oxford University Press. © Oxford University Press 2023.
DOI: 10.1093/oso/9780197689912.003.0002

Interpretation 1

It does not follow that because humans will explore space, space exploration is to be recommended. We are all going to die eventually, but many of us do not think highly of the prospect—inevitable outcomes are not necessarily good.

Interpretation 2

The claim that humans have a strong disposition or inclination to explore space does not fare much better. If the claim is true, and I suppose that it is, we may be interested to know why. But if we wish to know whether the exploration of space is wise, we have not yet moved from our starting point. After all, suppose that humans had a strong disposition to commit random murder. Should we condone random murder then? Should we yield to it or, if no longer able, encourage its exercise by the strong of body and heart? Surely not!

Appeals to nature or evolution do not get us very far.[3] For it would be a mistake to think that any trait with which nature has endowed our species cannot fail to be commendable. The trait in question might be an adaptation to an environment that no longer exists, and thus it may no longer serve us well. The fate of most species that ever lived has been marked by the development of characteristics well suited to one environment but woefully inadequate when the environment changed. Hence, they could become extinct. Imagine for the sake of argument that our species has a disposition toward war. In a nuclear-armed world, that is clearly a bad disposition to have. Thus, for the purposes of justification, it is not enough to show that a natural disposition exists: we must show that it is a *good* disposition.[4] Therefore, to justify space exploration, it does not suffice to learn that human nature tempts us to the heavens.

Interpretation 3

This third claim, that because of our nature we will go unfulfilled unless we explore space, is perhaps more promising. It would have to specify, however, in what important ways space exploration might fulfill us.[5] To become convincing, it would have to go beyond slogans and platitudes: it would have to supply strong evidence for the conclusion. It would require, that is, a good *argument* after all. This result thus returns us to the problem the supporters of space exploration face.

If those supporters hope to win over their fellow citizens, they should pay greater attention to why their opponents resist their entreaties. True argumentation cannot take place in a vacuum: you may not always be able to convince open-minded people, but you will improve your chances when you consider seriously what keeps these people from coming over to your side. This is why impatience tends to be rhetorically self-defeating. Supporters should start, then, with the objections critics offer against space exploration.

The Objections

Ideological Criticisms

The ideological critics argue that wisdom does not lie along any road that exploration may discover. "People," admonished one of their forerunners, the eighteenth-century French philosopher Jean-Jacques Rousseau, "know once and for all that nature wanted to keep you from being harmed by knowledge, just as a mother wrests a dangerous weapon from her child's hands; that all the secrets she hides from you are so many evils from which she protects you."[6] Rousseau's romanticism lives on among the ideological critics of space exploration. As they see it, the secrets our curiosity has pried from nature have brought us to the brink of disaster. We should have heeded Lao Tzu's warning: "[T]hose who would take the whole world to tinker with as they see fit . . . never succeed."[7] Ignoring this advice, Western science aims to control nature by interfering with it. Despite all the so-called progress of the scientific era, Western science has not succeeded and will not succeed.

To come to this conclusion, these critics argue, we only need observe the trends set in the previous century: the population explosion, the massive use of resources at an ever-increasing rate, and the unparalleled poisoning of the soil, the air, and the water of the Earth. It is doubtful that our planet can withstand this situation for long. Indeed, the Club of Rome Study, among others, has predicted a global environmental collapse around the middle of this century.[8] Even if this crisis does not spell doom for humankind—and it might—it deserves serious attention. The first thing we must determine is what makes all these dangerous trends possible. And, the ideological critics say, it does not take much to isolate the main factor: technology has coupled with the mentality of growth, and together they have run amok. But surely technology on such a grand scale could not have existed without prior great advances in science.

And what is space exploration, these critics ask, if not the expansion of this mentality of growth and scientific development? Hence, they find unacceptable the suggestion that space exploration can help us out of our dire straits. For that suggestion masks the imminence of the crisis and entreats us to engage in distracting pursuits—at a time when all our attention and effort should be concentrated on the abyss that is opening just a few careless steps ahead of us.

From this ideological perspective, space exploration is no more than another technological fix for problems that cry out for a different approach. The only solution is to realize that the crisis is upon us and to stop the activities that have created it. Above all, we must stop interfering with nature. Space exploration not only delays the real solution to the problem but is itself a symptom of the problem.

The ideological critics thus find little hope in the attempt to push science and technology beyond the confines of our natural habitat. Nor is the search for truth enough of a warrant. As Rousseau put it: "What dangers there are! What false paths when investigating the sciences! How many errors, a thousand times more dangerous than the truth is useful, must be surmounted in order to reach the truth? The disadvantage is evident, for falsity is susceptible of infinite combinations, whereas truth has only one form."[9] Wisdom dictates, then, neither investigation nor exploration, but living in harmony with nature.

An extreme fringe of ideological critics finds space exploration not just unwise, but positively evil. They fear, for instance, that a satanic science and technology will lead to the destruction of the human race at the hands of terrifying weapons. The Mercury Program may have sent astronauts on voyages of discovery, but its Atlas rockets also became the first intercontinental ballistic missiles (ICBMs) poised to destroy human lives by the hundreds of millions. Some writers, such as Lewis Mumford, hold that big science and technology magnify some of the worst human traits: not only have men wrought a brutal conquest of nature in "the effecting of all things possible" (in the words of that early promoter of science, the English philosopher Sir Francis Bacon), but the social "megamachine" they have produced has developed means for the complete extermination of the race.[10] Others like C. S. Lewis think that space exploration is a manifestation of unchecked pride and power. These critics argue that we have no right to pollute the heavens with our fallen race. Only a return to a more spiritual way of life can save us from a degrading future.[11]

How true another of Rousseau's dictums must seem to these critics: "Men are perverse; they would be even worse if they had the misfortune to be born learned."[12] If perchance there is a future, it will curse the day man reached for the stars.

In this day and age, however, neither this extreme view (which will be discussed in Chapter 9), nor the main ideological objection, comes as readily to mind as the following social objections.

Social Criticisms

The typical social critics oppose space exploration on what we may call humanitarian grounds. They cannot justify spending billions of dollars to find out what the Moon is made of at a time when hunger and poverty are rampant on our own planet. As they see it, space exploration takes money, resources, and talent away from helping people in need and from improving the quality of life for everybody. The human condition, one might quip, ought to take precedence over the condition of alien atmospheres and surfaces.

Unlike their ideological counterparts, some of these social critics may even think that space exploration is a good thing. They may agree that producing knowledge and satisfying our intellectual curiosity and thirst for adventure are all worthwhile goals. But they may also think that opera is a good thing without being prepared to spend billions of dollars in its support. The problem is not what space exploration tries to accomplish but rather the commitment of resources upon which other human needs may have a larger claim.

Now, when we speak of improving the human condition and satisfying other human needs, it is important to be specific. Presumably, there are areas of human suffering that come starkly to mind and demand immediate attention. Indeed, there are: millions, perhaps billions, of people in the world suffer from hunger and malnutrition; lack proper housing, education, and opportunity; and are afflicted by myriad diseases. But even this list does not convey the full extent of human suffering and misery. To do so, we must attempt to understand the hopelessness, the sense of being at the whim of tragedy—a tragedy brought about by nature, by humans, or by human indifference. These are the burdens we should lift from the people of the Earth before we go looking under the rocks of far-away worlds.

The two categories of criticism—social and ideological—are by no means clear-cut. Many critics would combine them or deviate in important respects from both. Still other critics may oppose space exploration on the prosaic grounds that it is not cost-effective.[13] Nevertheless, I think that my presentation of the main objections captures the essential challenge to the supporters of space exploration. Those supporters will, therefore, have to show, first, that in exploring the cosmos humankind is pursuing goals (or satisfying inclinations) that are in themselves worthwhile. And, second, that such pursuit does not proceed at the expense of even more worthwhile goals.

The Standard Case for Space Exploration

The supporters of space exploration do not feel overwhelmed by the challenge. They believe their case is straightforward: space exploration can contribute greatly to the reduction of human misery, the improvement of human life, and the preservation of the environment. In fact, it already has. To appreciate the actual and potential contributions of space, we need only pay attention to the function of satellites, the indirect consequences of space technology (spinoffs), and the opportunities that future exploration may create for humankind. And once we gain an appreciation of these contributions, we will have an answer to the social and ideological critics.

Satellites

Weather satellites have extended the range and accuracy of weather forecasts appreciably. The reason is simple: from space we see weather patterns that otherwise could be discerned only with great difficulty and never very accurately. Now we see them and track them.[14] Weather satellites warn us about freezes, hurricanes, and tornados, thereby saving crops, buildings, and many thousands of human lives.[15] And when disaster nonetheless strikes, communication satellites enable us to come to the assistance of people in peril or in need of relief.

Apart from this reduction in human misery, the drastic improvement in weather forecasting techniques is of great help to farmers in planting and harvesting, with obvious beneficial consequences for agriculture and the food supply of a hungry world.

Land satellites (LANDSATs and their descendants) are a useful complement to weather satellites. LANDSATs survey the Earth's resources from space, identifying minerals or types of vegetation by their responses to infrared, visible, or ultraviolet radiation. Often, a computer assigns contrasting colors (e.g., gold and purple) to slightly different frequencies (e.g., the frequencies of two closely related browns) that reflect from different materials (e.g., a mineral ore and dry vegetation). This use of "false color" and other computer tricks of remote sensing permit practically instant visualizations of the distribution of natural resources.

We can observe these patterns even on cloudy days, for we can take pictures at wavelengths of the electromagnetic spectrum not absorbed by water droplets. So we can reliably use LANDSATs to look for oil and other mineral deposits; make crop inventories; and carry out surveys of ice in lakes and of snow accumulation on mountains, thus helping to determine the likelihood of flooding. We can also measure the availability of water where it depends on snowmelt, estimate forest land, and determine the degree to which urban sprawl affects the surrounding environment. And finally, although the list could go on, we can often monitor the spread of pollution.

Since we can make estimates of the distribution of many resources, and of the productivity of many enterprises, we can see how space technology may be of great assistance in the fight against poverty. Remote sensing technology may also enable us to perform the inventories needed for the preservation of agriculture, wild lands, and wildlife. It seems then that land, weather, and communication satellites begin to answer the concerns of the social and ideological critics of space exploration (see Figure 2.1).

We must recall also the revolution in communications made possible by satellites. We now transmit information and contact people in ways that were unattainable prior to the launching of Sputnik I in 1957. Today, in the comfort of our living rooms, we can watch live on television a sporting or cultural event that is taking place on another continent, or use satellites, as in India, to bring education to large rural areas for the first time. All these changes in people's daily lives are mirrored by improvements in the practice of commerce, the gathering of news, and the relief of disaster.

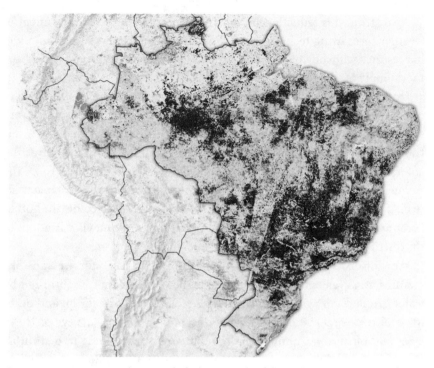

Figure 2.1. LANDSAT images help keep track of drought in Brazil (see color plate). (Image courtesy of NASA)

The Space Shuttle, as well as other piloted vehicles and the various kinds of space stations, complement these functions of satellites. A 1994 Space Shuttle flight yielded preliminary radar measurements of hitherto undiscovered structures around Angkor Wat, a famous archeological site in Cambodia. NASA's Jet Propulsion Laboratory then developed a sophisticated airborne radar system that allowed archeologist Elizabeth Moore and her team to discover four to six more temples and gain a better picture of the massive waterworks that were an integral part of the complex.[16]

Few of these accomplishments were likely through more conventional methods. Surveys from the ground could not compete with a perspective that permitted us to detect, at a glance, large patterns and to take inventories of minerals and vegetation. It might be imagined that perhaps airplanes could have flown high above the clouds to do a similar job for less money. But whereas satellites give us pictures of exactly the same spot time and again so we can make comparisons, the flight path of airplanes is never that precise. Nor are airplanes as reliable—they are subject to mechanical problems and the vagaries of weather. Moreover, it would have taken a fleet of thousands of airplanes to do what a single satellite does in passing over the Earth at its very high orbital speed. Using airplanes might have well cost us hundreds of times more, and the results would have been vastly inferior.[17] Today, we are beginning to use new generations of light planes and other flying devices to obtain more specific local information, normally interpreted in the larger context provided by satellite data.

In any event, many crucial jobs can be done only from space. For a variety of reasons, many weather and communication satellites must be placed exactly over the same spot on the Earth. For example, a satellite fixed overhead is extremely convenient, since we can then transmit and receive from it at any time. As the Earth rotates, the satellite must rotate with it so as never to lag or move ahead. Only a special orbit, called a geosynchronous orbit, 36,000 kilometers (22,000 miles) over the equator satisfies these requirements.

Of course, this is only the beginning. New generations of satellites will do far more. There is much progress, for example, in estimating crop yields using satellites. And five decades ago, the very idea of tracking devices for trucks and mountain climbers had an aura of science fiction. Three decades ago they were still a rarity. Future SEASATs and navigation satellites will not only survey the oceans but also contribute to the safety of travelers and cargo.[18] All in all, new kinds of satellites will improve in new ways the lives of billions and billions of human beings.

Space Technology and Economic Expansion

One of the most important aspects of space exploration, according to its supporters, is that the drive into space drives technology as well. This should be expected, they say, since in order to meet new challenges and solve new problems, we have to stretch our ingenuity well beyond the bounds of the ordinary. The result is beneficial because many of these advances in technology can be applied here on Earth. That is, from the space program we derive valuable "spinoffs." These come mainly in two categories. Some technological innovations are entirely extensions or applications of technology developed for space. And some others are developed independently of the space program but become well known, refined, or simply marketable because their use in the space program gives them a great boost.

The effectiveness of space technology in producing spinoffs cannot be determined precisely. One reason is that highly specialized technology may take a long time, often decades, getting to the marketplace. Penicillin and television, for example, were ignored for years before somebody decided to take advantage of them. Nevertheless, there appear to be direct links to the technology of space (particularly in the 1960s) in the development of new materials and techniques for aerodynamics, propulsion, electronics, and other fields. These developments, in turn, affected our systems of transportation, transmission of energy, and temperature control.

Even esoteric space technology often finds a home in the wider industrial world. The liquid hydrogen used as fuel in the Saturn V (the rocket that took men to the Moon) had to be kept at the incredibly cold temperature of minus 423 degrees Fahrenheit (–252 degrees Celsius). The fuel-tank insulation, which consisted of a one-inch thickness of polyurethane foam reinforced in three dimensions with fiberglass threads, is now applied in ships that transport liquefied natural gas. The conversion of the gas to liquid reduces its volume more than six hundred times, which makes it a far more economical and manageable cargo. But liquefied natural gas must be contained at about minus 260 degrees Fahrenheit (–162 degrees Celsius) to prevent loss by boil-off, a task moon technology has made safer and more efficient. Indeed, there are many applications of insulating materials designed for NASA. One such spinoff, Therm-O-Trol, provided the insulation required to keep the oil in the Alaska pipeline flowing at 180 degrees Fahrenheit (82 degrees Celsius). And Nunsun, a thin film of reflective insulation developed to protect spacecraft from intense solar radiation, can be sprayed on the windows of buildings to reduce the cost of cooling.

Examples of applications and their influence in industry and daily life multiply easily.[19] In the first two decades of exploration, space supporters pointed to that influence, whether direct or indirect, in thousands of products, from firefighting equipment and freeze-dried foods to hand-held calculators and digital watches. Indeed, the whole trend toward miniaturization, it is said, was spurred largely by the technical needs of the space program.[20]

Over time, of course, the list of products, and of the fields in which we can find them, grew much longer. Here is a small sample of applications and their origins in the space program.

Health and Medicine
- Nonsurgical breast biopsy system (space telescope technology: digital imaging)
- Ocular screening (NASA image processing), a photo-refractor that analyzes retinal reflexes
- Ultrasound skin damage assessment (NASA ultrasound technology)
- Voice-controlled wheelchair (NASA teleoperator and robot technology)
- Programmable Pacemaker (NASA computer technology)

Public Safety
- Emergency response robot used in hazardous duties (NASA robotics)
- Pen-sized personal alarm system (space telemetry technology)
- Self-righting life raft (Apollo program)

Transportation
- Advanced lubricants for railroad tracks, prevention of corrosion in electric plants, etc. (Space Shuttle Mobile Launcher Platform)
- Flywheel energy storage system, with fifty times more capacity than a standard car battery (NASA-sponsored studies)
- Studless winter tires (made from Viking Lander parachute materials)
- Improved aircraft wing and engine designs (from multiple NASA technologies)

These are examples chosen almost at random from among many thousands developed over the last several decades. I could have mention LASIK, artificial limbs, and countless others. NASA has a website and, since 1976, an annual journal, *Spinoff*, devoted to spinoffs.[21]

One could compile similar lists of space applications in other fields. Manufacturing, for example, benefits from NASA developments in magnetic liquids, new welding technology, and microlasers. An interesting spinoff is a system of magnetic bearings that allows motion of parts without friction or wear. This technology came from the Space Shuttle and is used for refining oil, building natural gas pipelines, and operating machine tools.

I have mentioned the origins of these spinoffs because the popular literature is full of questionable examples and some of the claims about the extent of space technology's influence on the development of specific products are disputed from time to time. Among the most notorious cases are Teflon, Velcro, ballpoint pens, and cardiac pacemakers. Carl Sagan recalls meeting the inventor of the cardiac pacemaker, "Who himself nearly had a coronary accident describing the injustice of what he perceived as NASA taking credit for his device."[22] NASA did design programmable versions of the device, such that a physician can make changes in an implanted pacemaker by communicating with it, making surgery unnecessary in many cases.[23]

Nonetheless, it seems that, as we saw earlier, space technology has improved considerably the quality of life for many people: here by saving it, there by making it more bearable,[24] and elsewhere by creating copious new opportunities in jobs and industries, or innovative products that enhance our work and our leisure. The enthusiasts suggest that much of this change is for the better, and that once people acknowledge the pervasive role of space exploration in their lives, they will realize that they cannot do without it.[25]

Economic Impact

In view of the foregoing description of the role of satellites and the impact of space technology on industry and daily life, we may suspect that space exploration contributes its fair share to the growth of the economy. Indeed, in the Golden Age of NASA (roughly from 1961 to 1976), space supporters claimed that exploration returns to the economy far more than it takes from it. This claim was backed by the flourishing of industries such as space communications, which employed tens of thousands and was soon worth tens of billions of dollars.[26] According to an econometric study, society's rate of return on NASA research and development (R&D) from 1960 to 1974 could be as high as 43%.[27] According to another, every dollar NASA spent on R&D would return seven dollars in gross national product (GNP) over a period of eighteen years (for an estimated social return rate of 33%).[28] Yet another concluded that the value of the technological acceleration NASA induced in gas turbines, computer simulation, integrated circuits, and cryogenic multilayer insulation might reach somewhere between $2.3 billion and $7.6 billion (in 1974 dollars, for the 1958–1974 period).[29] All these results could not fail to provide great ammunition against the social critics who doubted the cost-effectiveness of space exploration.

These considerations give supporters of exploration, at least in the United States, a chance to set straight a matter that irritates them greatly. They are irked by two assumptions that many social critics apparently make. The first is that, if the money were not spent for space, it would be used for fighting hunger and poverty instead. The second is that taking the money out of space programs and putting it into programs of direct aid would achieve the proclaimed humanitarian goals.

As these supporters see it, there is no guarantee that the first assumption would be borne out. Money may simply disappear from the budget without going to the poor and the hungry (or it may go to the military or to other government agencies). Moreover, instead of hampering those social programs, our adventures in space may have contributed to their growth. If we can put a man on the Moon, as the saying used to go, why can't we also do this or that? A nation inspired by the conquest of space was willing to undertake lofty goals of all kinds. Thus, supporters find very misleading the assumption that space exploration takes away from people in need.

As for the second assumption, we should consider that at its highest level of funding, in 1966, the budget for space exploration amounted to 0.8% of GNP ($5.9 billion); by 1979 the space budget had dropped to 0.2% of GNP.[30] In 2009, NASA's budget was $17.8 billion, or 0.5% of the U.S. budget,[31] a slightly larger percentage than 2020 (0.48%).[32] It seems unlikely that such a relatively small percentage would make a significant difference in tackling problems that budgets many times larger do not solve (e.g., expenditures for anti-poverty programs in the late sixties were six times higher than those of the space program—they are far more than that now).[33] Furthermore, if we are concerned with fighting poverty, it seems at least sensible to encourage programs that pay for themselves while increasing productivity and economic growth.[34]

Some Reservations about the Economic Case

The enthusiasm for the economic case has waned since the Golden Age of exploration, in great part because the Space Shuttle made it too expensive to place things in orbit. It has been said that if the alchemist dream of the "philosopher's stone" (to turn lead into gold) could be realized simply by taking the lead aboard the Shuttle, it would have cost more than just buying the gold!

To make matters worse, the Shuttle was not exactly a model of reliability. After the Challenger blew up in 1986, the United States began to use rockets regularly again. The Russian, European, Japanese, and Chinese space agencies, of course, use rockets also,[35] but although rockets are cheaper than the Shuttle, they are still expensive ($10,000 per kilogram, as of 1995).[36] However, with the SpaceX Falcon 9, the cost has gone down to just $2,720 per kilogram, as delivered to the International Space Station. Quite a few other private companies have joined the commercialization of space, which not only lowers prices but gives access to a variety of technologies. They include Blue Origin, Boing, Paragon Space Development Corporation, Sierra Nevada, Corporation, United Launch Alliance, and Orbital Sciences Corporation. Apart from them, many countries are developing their own space capabilities. These are reasons to feel optimistic about the Renaissance of space exploration mentioned in Chapter 1. Incidentally, as we will see in Chapter 7, one of the main disappointments of the Space Shuttle is that it had been billed as the inexpensive option because it was partially reusable.

Even during the Golden Age, the economic studies mentioned earlier might have been too optimistic. Some of them were based on assumptions about the general relationship between R&D and economic growth, with the expenditures for space technology plugged in—assumptions not universally accepted by economists. And studies that try to account for the specific influence of space technology on a wide collection of industries must surmount serious difficulties. The first difficulty is that knowledge is the most common byproduct of space exploration. It is difficult to quantify knowledge, and even more difficult to trace precisely how it affects the economy as a whole. A few examples of such effects can be given here and there, but a comprehensive account is a great challenge.[37]

The second difficulty is that space hardware's effects on the economy are hard to trace because there is often a considerable lag between invention and assimilation, as it happened in the cases of television and penicillin.[38] Moreover, the invention may undergo a series of transformations that are influenced by many factors, including other inventions from completely different fields, or the presence of special social and economic conditions. Catalytic converters to reduce automobile pollution, for example, depended for their acceptance on strong environmental activism in North America. In response to this activism, governments came to support the development of unleaded gasoline and passed laws against engines that used leaded gasoline. It also made automobiles more expensive. In many poor countries of the Third World, where the economic conditions were harsher, catalytic converters were a rarity.

Separating all these factors and settling all these issues is necessary before one can offer truly solid numbers to support the contention that space is a better investment than others that society may contemplate. Thus, in spite of all the money figures thrown around, with a few important exceptions such as telecommunications and navigation, the economic case is mainly qualitative,[39] even if some find it very suggestive.

However, as faulty as the econometric and comparative studies may be, the space enthusiasts can find solace in the realization that once an aspect of space exploration is commercialized, its economic impact may be considerable. In 2001, space *commercial* revenues worldwide reached about $83 billion[40] and had almost tripled by 2006.[41] This sum by itself, however, does not reveal the economic growth it spurs in many other industries. It seems that, after all, the onus is on the social critics to explain why the economic justification of space falls short of the mark.[42]

Exploration and Future Opportunity

The appraisal of how much space exploration has done for us pales by comparison with the appraisal of how much more it may do in the years to come. The change of perspective is significant: whereas until now we have only tried to reach outer space and survive there for short periods, we will soon be in a position to live in space, industrialize it, and really put it to work for our benefit. Space presumably has two main advantages for industrialization: low gravity and a nearly perfect vacuum. These two advantages combined can bring us a treasure of new materials, including metal alloys, super-crystals, and extremely pure semiconductors and pharmaceuticals.

Consider the technological promise of low gravity ("microgravity" in the jargon of the trade). Under the influence of gravity, objects have weight. When we mix different substances, gravity pulls the heavier to the bottom and leaves the lighter on top. In a similar fashion gravity creates openings between molecules—openings that allow impurities into the mix. Remove gravity and we can mix the substances evenly and without impurities.

The prospects for new technologies dependent on microgravity are said to be very encouraging. One of those technologies is levitation melting, in which molten metals can solidify without the use of a container (further reducing the problem of impurities). By injecting gases into the heated mixtures, we can produce alloys that are not possible on Earth. Some of those alloys may have extraordinary properties; we may, for example, produce a form of steel as light as balsa wood. In medicine, the new purification techniques may be valuable in the investigation of new drugs or in the mass production of some drugs that are currently too expensive to manufacture.

Some of these possible new products would have to be manufactured in space, but others could be developed in space and then made on the planet. Once the feasibility and practicality of these products has been demonstrated through space research, earthbound industry would be more willing to get around the obstacles that gravity presents to their manufacture down here. The vacuum of space combines with microgravity to provide further opportunity for this sort of industrial research in metallurgy, thin-film coating, and welding, among others.

I must point out, however, that these exciting possibilities have been proclaimed almost from the beginning of the space program. It is at least worrisome that sixty years later industrialists do not yet seem to be flocking to take advantage of them. Part of the problem may well be that the Space Shuttle, instead of reducing launching costs—which was the main purpose for building it—increased them dramatically. We will see later to what extent the International Space Station has fulfilled this promise.

If the costs of transportation can be reduced, setting up factories in space may have several advantages. Large structures can be built in space without many of the problems of foundation and support that gravity forces us to solve down on Earth. Without atmosphere, to say nothing of bad weather and pollution, machines can work for extremely long periods of time. And the energy they require can be obtained cleanly and efficiently from the Sun.

All the industrial and technological advantages mentioned so far suggest how space exploration may play a major part in solving some of the most urgent problems of the Earth. Our world faces a double jeopardy: increasing demand for energy and dwindling of resources. In trying to obtain more energy, we use up even more resources and, to make matters worse, produce greater amounts of pollution, which in turn affects some of our other resources, as well as our health and general well-being. For example, fossil fuels are the usual source of industrial energy. As we use them, we release ever-greater amounts of carbon dioxide (CO_2) into the atmosphere. According to the prevailing view, as CO_2 increases, the resulting greenhouse effect might raise the temperature of the planet enough to change the climate and melt much of the water now frozen in the polar caps.[43] In the worst-case scenario, large areas millions of humans inhabit will be flooded out of existence.[44]

To forestall these dire consequences (which I will discuss in Chapter 4), supporters of exploration have made proposals that range from the building of solar power satellites to the mining of the Moon, the asteroids, and eventually other planets. In 1968, Peter Glaser proposed a solar power satellite to collect sunlight, transform it into microwave energy, and beam that energy down to Earth.[45] In space, sunlight is plentiful and likely to last for billions of years; solar power satellites release no CO_2; and environmental studies indicate that beaming this energy as nonpulsated microwave beams would be less harmful to plant and animal life than the existing alternatives. In the early proposals, one solar power satellite the size of Manhattan would provide as much power as ten nuclear power plants without the attendant risks of radioactive leaks and meltdowns.[46] In our time, with advances in photovoltaics (e.g., solar cells) and other fields, a collector about the size of half a football field might be able to produce one megawatt of power. Other space exploration supporters have suggested moving some of the most polluting industries to space. The promise of space exploration is then very enticing: abundant energy and a safer, cleaner environment.

Critics of these proposals have argued that the mining of the enormous quantity of materials required to build such structures would cause major environmental headaches, while the many thousands of flights by giant rockets to haul the materials into orbit might damage the atmosphere and are certain to cost far too much—in the hundreds of billions of dollars—at least for the system as presented to the U.S. Congress in the late 1970s. Congress found the proposal technologically feasible but accepted the criticisms and refused funding.

These criticisms seemed misleading at the time. The late physicist Gerard O'Neill, one of the most vocal proponents of the idea, had said all along that most of the required materials (e.g., aluminum, oxygen, and silicon) could be rather easily extracted from the Moon, placed in lunar orbit, and processed there.[47] The gravity pull of the Moon is only one-sixth that of Earth, and thus the materials could be shot into lunar orbit, at great savings of energy and money, by what O'Neill called "mass drivers": long superconducting rails that would use powerful electromagnetic fields to accelerate metal buckets full of lunar soil.

This project would be the beginning of the eventual colonization of the solar system, for no insurmountable technological barriers would then keep us from the abundant resources available in the asteroids, nor from building large habitats in space. To paraphrase O'Neill, the closing of the Earthly frontiers would be compensated for by the opening of the "high frontier" to the needs and hopes of humankind.

Whether projects of such magnitude are truly feasible in the next few decades remains a matter of controversy, while the enthusiasm for building O'Neill's cities in the Lagrangian points between our planet and the Moon seems to have dissipated.[48] A sobering sense of reality developed in the late 1980s when people realized that the Shuttle would never be the transportation system that O'Neill had assumed. Instead of fifty inexpensive flights a year, we were lucky to get five, and at astronomical costs (pun intended). This was no system for colonizing and mining the Moon. More recent proposals for solar power satellites suggest much smaller projects, though still large, for considerably less money, even though all materials would come from Earth.[49]

To summarize, from the supporters of exploration we get an impression of great accomplishments in the past and even greater possibilities in the future. Their case, which by now is pretty much standard in the pro-exploration literature, seems quite impressive. It points out to social critics that space exploration reduces human misery and improves life on Earth. It tells ideological critics that space technology helps in controlling pollution and in monitoring the environment as a whole; and to both it promises that the new coming Golden Age of space exploration will do much to solve some of our most serious problems. Let us review it in outline:

(1) Satellites:
    (i) Weather:
        (a) Save lives
        (b) Help agriculture
        (c) Help transportation
    (ii) Sea:
        (a) Find resources
        (b) Tell us about environmental impact
    (iii) Land:
        (a) Find resources
        (b) Tell us about environmental impact
    (iv) Communications:
        (a) Help commerce
        (b) Make our lives easier
(2) Spinoffs:
    (i) New technologies
    (ii) New economic opportunities

(3) Future developments:
  (i) More of the same, many new things, and all on a much grander
      scale

So what is wrong with this standard case? Supporters feel that the critics
have received more than they had a right to demand—that they are looking
a gift horse in the mouth and turning it down after finding his teeth in excellent condition. But lest we be too hasty in dismissing the critics, we should
consider whether the spirit of their objections has been met.

New Round of Objections

The Ideological Critics' Reply

A justification that involves technological and economic growth is not likely
to impress ideological critics. Indeed, they see the alleged benefits as causes
for concern. For many of these critics, and especially for some influenced
by the environmentalist movement, the very idea of space exploration is
not only unwise but also immoral. They are particularly harsh to some of
the grandiose proposals for going into outer space to solve pressing terrestrial problems. According to Wendell Berry, for example, the lesson that we
should learn from the closing of the earthly frontiers "calls for an authentic
series of changes in the human character and community that, if made, will
afford us the spiritual resources to live both within our material means and
with each other."[50]

Space exploration, he thinks, tries to outflank the lesson entirely. The
space enthusiast—and here Berry has Gerard O'Neill in mind—ignores what
is essentially a moral problem (i.e., the changing of human character and
community) and offers technological solutions instead. The morality of the
space enthusiast is thus both shallow and gullible, for he offers "a solution to
moral problems that contemplates no moral change."[51] Space exploration, to
someone like Berry, could only be "a desperate attempt to revitalize the thug
morality of the technological specialist, by which we blandly assume that we
must do anything whatever that we can do."[52] According to another critic,
Dennis Meadows, "What is needed to solve these problems on earth is different values and institutions—a better attitude towards equity, a loss of the
growth ethic. . . . I would rather work at the problems here."[53]

At first sight Berry seems to beg the question. According to him, the closing of the earthly frontiers presents a moral problem to which only moral solutions are applicable. Gerard O'Neill and other space enthusiasts ignore the moral problem. Thus, Berry concludes, the space enthusiasts are not only doomed to failure but are also immoral (not just mistaken or unperceptive). But what O'Neill and the others question is precisely whether all the frontiers have in fact closed. And certainly, if those frontiers haven't closed, we have no reason to believe that we face that moral problem. In assuming that the high frontier is not a genuine option, Berry heaps moral blame on the space enthusiasts while begging the issue in question.[54]

But perhaps there is a more sympathetic reading of Berry's position. What he may have in mind is that the experience of the (partial) closing of the earthly frontiers is enough to show that the Western approach to nature is inherently unwise, and thus that its extension through space exploration is destined to fail. On what grounds should we trust O'Neill's grandiose plans for gigantic solar power satellites, let alone those for artificial worlds (his space colonies)? Surely projects of such magnitude cannot be made plausible by mere theoretical proposals. How can we be assured that no essential detail has been left out?[55] The most straightforward way to resolve this issue might be to demonstrate the feasibility of increasingly more complex stages of these projects. O'Neill would have been quite agreeable to this suggestion, but Berry and many other ideological critics would probably resist it. The reason for resisting it is that to undertake such demonstrations we first need a large commitment to space exploration, for the demonstrations require that we build and operate very large structures in space. But given the poor record of big technology, Berry would say, why should we extend it the benefit of the doubt on such a scale?

The ideological critics are thus not impressed by the suggestion that space exploration can help correct some of the excesses and mishaps of technological civilization. Nor are they impressed by the claim that space exploration enables us to appraise better how critical our environmental situation is. Giving credit to space exploration in this regard may call to their minds the case of a drunk who drives his car into a bed of flowers. Should credit go to Detroit for inventing the tow truck that gets the drunk's car out? Such would be the wrong approach to the problem. What we need to do is prevent the situation in the first place. Space is, then, a delusion, for it offers more growth and technology to stop the mess caused by growth and technology. Of course, the more we foul up the world, the more space will look like a necessity. But this is a false technological panacea. It is rather like a pain reliever that keeps the patient from having the operation that will save his life. As Wilson Clark puts it, "[O'Neill] speaks in terms of a 'first beachhead in space,' evoking the image of greener grass on yonder hill. Unfortunately, we have little time in which to prevent the elimination of the vegetation altogether."[56]

The urgency of the situation, as these ideological critics perceive it, makes unwarranted our engaging in any more technological detours. Western man's approach has brought the world to the edge of crisis by marrying technology to the mentality of growth. This ideological criticism touches the heart of space exploration insofar as science is supposed to provide the promissory note that underwrites that marriage in the first place. Once again, the satisfaction of scientific curiosity—at least where "big science" is concerned—may be seen as a disturbance, an interference with nature. The emphasis on beneficial results is only a smoke screen: in the long run only, a change of attitude can be beneficial. Anything not in harmony with nature is bound to make us fail. In the eyes of the ideological critics, space exploration amounts to a distraction at a time of crisis—the siren voice that calls us from the cosmos still sings the tune of our doom.

I will offer three comments on this controversy. First, most of the vehemence against O'Neill was caused by his suggestion to build space colonies, some of which would house millions of human beings (see Figure 2.2).

The idea that one could build artificial self-sufficient environments on that scale seemed naive and arrogant to his critics. As the many difficulties encountered in trying to create such a closed environment in Biosphere 2 indicate, we are a long way from knowing enough to attempt anything remotely approaching the ambition of O'Neill's projects.

Biosphere 2, a three-acre compound in the Arizona desert, was originally designed to prepare future space colonists by having them live sealed off from the rest of the world in a self-contained environment for long periods of time. The first attempt failed: crops were poor, oxygen fell to a dangerous level (15%), and there were several violations of the planned isolation. The second attempt was aborted. Eventually the facility was turned over to a team from Columbia University to perform environmental experiments, many of them connected to the ways buildup of CO_2 affects a variety of habitats.[57] Today the facility, now operated by the University of Arizona, offers the opportunity to do valuable research on a variety of environmental questions. Indeed, Biosphere 2 may contribute to the realization of O'Neill's dream someday, but not soon. In the meantime, it is clear that the ideological critics' warnings

Figure 2.2. A version of O'Neill's Island One, based on a Bernal Sphere. Rings on top of sphere would be individual torus areas probably devoted to agriculture. Larger versions could house millions of humans (see color plate). (Image courtesy of NASA)

were not entirely off the mark. A peculiar consequence of the scientific approach to Biosphere 2 is that environmentalists and supporters of exploration have found common ground. Thus, we now find organizations such as The Earth and Space Foundation, of which Charles Cockell is the Chair, devoted to bridging environmental concerns and space exploration along the lines advocated in the rest of this book, particularly in Chapters 4 and 10.[58] And even more recent work defends the notion that the two movements can go hand in hand.[59]

Now that the size of the possible space solar collectors has been reduced to that of a football field, O'Neill's proposal to bring clean power from space should sound far more amenable to those with environmental concerns.[60]

Third, space enthusiasts often present solar power satellites as the main scientific alternative to the energy crisis. But other scientific proposals may serve us just as well, if not better. For example, Roland Winston and others have demonstrated that by keeping light from forming images (nonimaging optics), it is possible to achieve here on Earth temperatures much higher than those on the surface of the Sun. Nonimaging optics may also be used to power lasers and even spacecraft. At the moment, most of the applications are in the heating of buildings and the like, but with the advent of the right kind of photovoltaics, it will be possible to transform that energy into electricity.

If that happens, we could have a revolution in electrical power plants analogous to that brought about by personal computers in information. Personal computers liberated us from the institutional giant computers of four decades ago. Nonimaging power generators would perhaps liberate us from giant power plants—for a lot less money and at far less risk. Every building would have its own extremely efficient, nonpolluting, and independent means of generating all the electrical power (as well as heat and air-conditioning) it needs. Power cables to housing areas would become a thing of the past. Of course, this particular technology may not pan out any better than solar power satellites, but its very possibility should make us beware of making space technology the only scientific alternative.[61]

Solar power satellites are not even the only alternative space science and technology suggest. Jerry Kulcinski and John Santarius claim that a deuterium-helium-3 reactor would offer abundant, cheap energy free of radioactive waste. Deuterium is an isotope of hydrogen, and it is not difficult to get, but there is no helium-3 on our planet. We could mine it on the Moon, though, and, Robert Zubrin adds, we could also scoop up large quantities of it in the atmospheres of Jupiter and the other gas giants of the solar system.[62]

Of course, this proposal comes, as all do, with several ifs attached (if fusion can really be made to work, if we can really mine helium-3, etc.), as do the other proposals to solve our energy problems by going into space.

In the meantime, the ideological critics impatiently point to solutions that, they believe, truly get to the heart of our planet's problems.

The Social Critics' Reply

Despite the long list of actual and potential benefits of space, social critics find that the standard case for exploration falls short of its target. For many important space activities do not have the obvious beneficial consequences of weather and communication satellites. Where is the obvious payoff from a probe of Jupiter or Titan, from landing a vehicle on Mars, or from scooping a bit of Halley's Comet? Few accomplishments of space exploration rank as high as the discoveries made with telescopes in orbit. But how is the information from space astronomy going to put food in children's mouths or a roof over their heads?

In emphasizing the practicality of space technology, the standard case makes intellectual orphans of the very things that bring to exploration an air of mystery and excitement. What it leaves out is the heart of space exploration: our sense of adventure, our urge to explore, our need to satisfy our curiosity. A justification along practical lines fails because it excludes those aspects of the enterprise that ignite the imagination and stir the soul about the conquest of the cosmos.

Many supporters of space exploration would like to argue at this point that scientific knowledge has value in itself, but this only brings us back to the original debate. Is scientific knowledge more valuable than achieving this or that social aim? We have not answered that question yet. Of course, merely a small portion of the space budget is allocated to science (while most of it presumably goes for more obviously practical activities), and since the space budget is not that large to begin with, taking the money away from the heart of space exploration is not going to solve the social problems anyway. Nevertheless, many social critics would not accept this reply because the actual sums spent on space science are large, even if they represent a small fraction of the space budget in the United States. The proposed price tag for the Hubble space telescope alone was around $1.5 billion, and the actual costs so far are of the order of $12 billion. The new James Webb telescope has just gone into space, after cost overruns of $10 billion. Operations would add much more over the years to come. That kind of money will not solve all the social problems of the world, but the social critics think that its judicious investment may do a lot of good. Even more convincing: even though many space scientists opposed the building of the space station (as we will see in Chapter 7) the cost so far, including overruns, is over $150 billion and mounting!

Serendipity

Supporters may respond that a crucial aspect of their case has not been presented adequately. All those benefits they proudly mention are the results of having yielded earlier to the call of the heavens. When humans first explored, we did not know for certain that so many good consequences would repay our efforts—very often we had no inkling. The pursuit of scientific exploration pays because of the *serendipity* of science; that is, because of the unintended benefits that science yields. This realization, supporters think, should make us share their faith in the future of exploration and believe with them in the continuous flow of treasure from our spaceships, even when they cannot say what that treasure will be.

The critics, however, may doubt that the prior performance of the space program is enough warrant for that faith. Having gotten water out of a well before does not guarantee an inexhaustible supply. Even space activities near Earth, which are often beneficial because of the vantage point they provide, are beginning to experience problems of saturation. Geosynchronous orbit, for example, is becoming crowded with communication satellites that are beginning to interfere with each other. Moreover, space debris—mostly from the breakup of rockets—is becoming a hazard to operations in lower orbits.[63] And now Elon Musk's Space X is launching small satellites into low Earth orbit that interfere with astronomical observations. Advances in technology will probably solve these problems, but we still can see that linear growth of benefits is not automatic.

Furthermore, the evidence for serendipity becomes more tenuous the farther we go away from Earth. Critics may wonder what link exists between a probe of Jupiter's atmosphere and the lot of those who breathe Earth's atmosphere. Moreover, although the history of science offers some striking instances of serendipity—for example, the nineteenth-century Scottish physicist James Clerk Maxwell's research on electromagnetism made possible television and computers, two inventions which Maxwell himself could not have foreseen—anecdotes make for a very one-sided historical analysis, for little is ever said about the overwhelming majority of the research carried out during the nineteenth century. Did all that science yield practical benefits, or only the most exceptional science, as Maxwell's surely was?

Even if critics grant that there is a strong connection between exceptional science and serendipity, supporters still have to show that the research they propose will prove to be exceptional. Or else they have to show that serendipity is a feature of most science. If they cannot show either, their standard case will have the ironic consequence of exposing the heart of space exploration to the narrow-minded whims of cost-benefit analysis. That is hardly the stuff dreams are made of.[64]

Furthermore, as far as many social critics are concerned, there is another serious objection: if spinoffs are so valuable, does it not make more sense to spend the money directly in the relevant fields?

The Supporters' Next Move

How could the supporters begin to address these objections? They need an argument to show that, because of its nature, scientific exploration makes serendipity somehow inevitable. Does such an argument exist? It does. I will provide it in the following chapter and defend it in the rest of the book. In that context it will be easier to appreciate the force of the additional concerns supporters have about protecting humanity from the impact of asteroids and from the eventual turning of the Sun into a red giant.

Notes

1. Arthur C. Clarke, "The Challenge of the Spaceship," *Journal of the British Interplanetary Society* (December 1946), p. 68. With great honesty he wrote: "Any 'reasons' we may give for wanting to cross space are afterthoughts, excuses tacked on because we feel we ought, rationally, to have them. They are true but superfluous— except for the practical value they may have when we try to enlist the support of those who may not share our particular enthusiasm for astronautics. . . . The search for knowledge, said a modern Chinese philosopher, is a form of play. Very well: we want to play with spaceships."

2. As quoted in T. Greve, F. Lied, and E. Tandberg (eds.), *The Impact of Space Science on Mankind* (Plenum Press, 1976), p. 13.

3. Actually, if most men, let alone all men, had a disposition to rape, we might suspect that they had it by virtue of being (male) humans. This, of course, would not justify it.

4. Some philosophers may be disappointed by my failure to accuse these supporters of exploration of committing the "naturalistic fallacy." In saying that since exploring is part of our nature we need no further justification, the supporters might be thought to claim that "being in one's nature" is sufficient justification to show that exploration is a good thing. And this claim the philosophers would interpret as deducing values from facts (or prescriptions from descriptions, or an "ought" from an "is")—a deduction that presumably amounts to a fallacious inference: the naturalistic fallacy. Some ideological critics also can be thought to commit the fallacy when they charge that the practice of science disrupts the harmony of nature. I am not quick to make such accusations, however, because they fit only the most simplistic and often distorted version of the position under discussion. My impression is that most of the "classic" examples of the naturalistic fallacy are nothing but arguments that were taken out of context, oversimplified, and distorted so as to serve as illustrations of a neat "logical" point. A more detailed discussion of this claim belongs in a different kind of work (see my "Review of Peter Singer's *The Expanding Circle*," *Explorations in Knowledge* [Spring 1987], p. 43; "Evolution and Justification," *The Monist* 71, no. 3 [1988]; and especially "The Morality of Rational Ants," Ch. 11 of my *Evolution and the Naked Truth*

[Ashgate, 1998]). At any rate, we can see in the text that follows that the supporters of exploration can, with some thought, express their insight in terms of fulfillment—a move that does not strike me as fallacious.

5. I am sympathetic to this intuition, as will be seen throughout this book, although I am critical of some of the ideas involved in what some have called the "space imperative," which is discussed by G. Genta and M. Rycroft in their book *Space, the Final Frontier?* (Cambridge University Press, 2003). Indeed, some versions of the space imperative fall under the interpretations 1 and 2 criticized earlier.

6. Jean-Jacques Rousseau, "Discourse on the Sciences and the Arts," in *The First and Second Discourses*, trans. Roger and Judith Masters (St. Martin's Press, 1964), p. 47.

7. As quoted by E. F. Schumacher in *Space Colonies*, a Co-Evolution Book, ed. S. Brand (Penguin Books, 1977), p. 38.

8. The Club of Rome's main report was *The Limits to Growth* (Signet Books, 1972).

9. Rousseau, "Discourse on the Sciences and the Arts," p. 49.

10. See particularly Lewis Mumford, *The Myth of the Machine II: The Pentagon of Power* (Harcourt Brace Jovanovich, 1970).

11. C. S. Lewis, *That Hideous Strength* (MacMillan, 1967).

12. Lewis, *That Hideous Strength*.

13. This kind of objection is found mostly among politicians. Former Senator Proxmire was perhaps the clearest example. A more theoretical objection on political grounds is offered by those who oppose space exploration, not because they think that some other goals are more worthwhile, but because they think that the state has no right to tax citizens in order to engage in such projects. They find the question of worth simply irrelevant. In answer to these libertarian concerns, I can always point out that my aim is to determine if space exploration is worth supporting or not, whether the support is provided by the state or by the free initiative of citizens. Nevertheless if one assumes that only the state can bring about the exploration of space, some interesting political consequences should be considered. One line of thought is that it will become clear that the state has a legitimate role to play in science. Against this it may be argued instead that a massive role by the state in space exploration distorts both science and society. A most intriguing exploration of these issues can be found in Walter McDougall's *The Heavens and the Earth: A Political History of the Space Age* (Basic Books, 1984).

14. Until 1980 weather satellites led to significant improvements in weather predictions mostly in the Southern Hemisphere and over the oceans, not in the advanced, populated areas of the Northern Hemisphere. The reason is that temperature and pressure at different altitudes could be better determined by other technological means. And improvement in the computer weather programs in 1980 led to new and more powerful forecasting techniques in which satellites played a crucial role. Future generations of weather satellites will provide more refined measurements.

15. The number of hurricane casualties shows a steady decline. In the 1900 Galveston Hurricane, for example, about 8,000 people died. Death tolls around 1,000 resulted from hurricanes in 1919 and 1926. It was not uncommon to see even higher casualties in the first part of the twentieth century. By the 1960s and 1970s, the numbers were

more typically in the low hundreds. Today they are in the dozens. The one disastrous exception is Hurricane Katrina, which may have caused close to a thousand deaths in Louisiana, Mississippi, and Alabama. Here we have a case, however, in which the warning was delivered but not properly heeded. In many cases, it seems, the reduction of casualties over the last two decades made some people overconfident, and many refused to leave the area. To make matters worse, the evacuation plan for the city of New Orleans was not followed, even though a run-through a year earlier showed that more than 100,000 people were likely to stay in the city unless city and school buses (and probably additional transportation from the state of Louisiana) were pressed into service. For over twenty years it was known that a hurricane that strong would destroy the levees and flood the city. But no one took steps to prevent the calamity. A tragedy of errors turned a serious but still manageable problem into probably the worst natural disaster in the history of the country. One shudders at the thought of what would have happened without the satellite warnings. Incidentally, military and civilian satellites, including Ikonos, a private imaging satellite operated by Space Imaging, are already giving us a reliable assessment of the damage.

16. *Science News* 153 (February 21, 1998), p. 117.
17. For details, see Greve et al. (eds.), *The Impact of Space Science on Mankind*, p. 82. For a summary of the benefits derived from LANDSATs and environmental satellites, see the same work, pp. 67–111. In it there are also discussions of communication and weather satellites.
18. Satellites may also be used to survey the resources of the oceans and to carry out sophisticated measurements of temperature and height of the ocean waters.
19. Literature describing actual and possible applications of space technology is easily available at any bookstore. Apart from this popular literature, the reader may wish to consult NASA's periodic summaries, appropriately entitled *Spinoff* (many of the examples given in this chapter are taken from *Spinoff 1979* and *Spinoff 1984*). Of almost historical interest in the forecasting of the industrial benefits of space exploration is Neil P. Ruzic's *The Case for Going to the Moon* (Putnam's Sons, 1965).
20. Although, as Jerome Schnee points out, the contributions of defense R&D were also very large. See his "The Economic Impacts of the U.S. Space Program," in T. Stephen Cheston, Charles M. Chafer, and Sallie Birket Chafer, *Social Sciences and Space Exploration* (NASAEP-192, 1984), p. 24.
21. NASA Spinoff: Home. https://spinoff.nasa.gov/
22. Carl Sagan, *Pale Blue Dot: A Vision of the Human Future in Space* (Random House, 1994), p. 272.
23. Programmable Pacemaker. https://spinoff.nasa.gov/spinoff1996/25.html
24. For medical advances produced by the early exploration of space, see T. E. Bell, "Technologies for the Handicapped and the Aged," NASA Technology Transfer Division, 1979, a report for the Select Committee on Aging and the Committee on Science and Technology, U.S. House of Representatives.
25. For an account of the accomplishments of the American space program during its Golden Age, see F. W. Anderson, Jr., *Orders of Magnitude: A History of NACA and NASA, 1915–1976* (National Aeronautics and Space Administration, 1976).

26. For a brief but comprehensive account of the economics of the space program, see Schnee, "The Economic Impacts of the U.S. Space Program." The following works are of interest for a variety of reasons: Mary Holman, *The Political Economy of the Space Program* (Pacific Books, 1974); Bernard Lovell, *The Origins and International Economics of Space Exploration* (University Press of Edinburgh, 1973); Nathan C. Goldman, "Space Race: The U.S. Won the Sprint, Can We Compete in the Marathon?" (unpublished).

27. Chase Econometric Associates, Inc., "The Economic Impact of NASA R&D Spending: Preliminary Executive Summary" (NASA-2741, 1975). This and additional references can be found in Schnee, "The Economic Impacts of the U.S. Space Program." The numbers, though, are peculiarly close to those given by Edward Denison, "Accounting for United States Economic Growth, 1929–69" (Brookings Institution, 1964), according to which technological innovation accounted for 44% of U.S. productivity increases for that period.

28. Midwest Research Institute, *Economic Impact of Stimulated Technological Activity* (1971).

29. See the Appendix to Schnee, "The Economic Impacts of the U.S. Space Program," p. 96.

30. These figures come from *Economic Reports of the President*. They are discussed in connection with the issues of this section by Mary A. Holman and Theodore Suranyi-Unger, Jr. in "The Political Economy of American Astronautics," AAS Paper No. 80-51, March 1980, p. 41.

31. Office of Management and Budget, "Budget of the United States Government, Fiscal Year 2011" (http://www.whitehouse.gov/sites/defaultlfiles/omblbudget/fy2011!assets lbudget.pdf), pp. 132 and 146.

32. *Budget of NASA-Wikipedia*. https://en.wikipedia.org/wiki/Budget_of_NASA

33. Holman, *The Political Economy of the Space Program*. "Between 1964 and 1968," she writes, "federal aid to reduce poverty rose from an annual expenditure of $13 billion to $24 billion. During that period, spending for the exploration of space dropped by about $300 million" (p. 347). And this was during the Apollo program—what we may call the Golden Age of space. Furthermore, she argues that whereas the space programs normally achieved their objectives, the programs designed to combat the causes of poverty generally did not. Eventually, Holman believes, those social programs came to be viewed as "pure welfare expenditures rather than investments that ultimately result in productivity gains." It makes no sense, then, to divert resources from programs that do increase productivity, such as the space program. In Holman's words: "Productivity gains contribute to economic growth, economic growth contributes to higher levels of national income, which, in turn, provides a wider tax base to support larger income and consumption maintenance poverty programs" (p. 348).

34. This line of argument will be revisited toward the end of the chapter.

35. They are being joined by Brazil and several other countries.

36. For a spirited discussion of these matters, read R. Zubrin, *Entering Space: Creating a Spacefaring Civilization* (Tarcher/Putnam, 1999), Ch. 2. His source for the cost/

kilogram is S. Isakowitz, *Space Launch Systems* (American Institute of Aeronautics and Astronautics, 1995).

37. Holman discusses this point in detail. NASA acknowledges it also, as can be appreciated in the agency's response to a critique of the Chase Econometrics study by the Government Accounting Office (GAO). In Holman and Suranyi-Unger's words, "NASA simply stated that the GAO results showed that because empirical measurement in economics is an inexact science, *ranges rather than absolute magnitudes are important.* [My emphasis.] That is how NASA justified its claim that the GAO findings [that the NASA R&D rate of return was about 25% to 28%, instead of the 43% claimed by the Chase Econometrics study] in fact, reinforced the results of the Chase Study." Holman and Suranyi-Unger, "The Political Economy of American Astronautics."

38. For a fascinating discussion of the relationship between "leap-frogging" technological innovations and economic growth, see Arthur M. Diamond Jr., "Schumpeter's Creative Destruction: A Review of the Evidence," *Journal of Private Enterprise* XXII, no. 1 (2006): 120–146. See also "Economics of Science," in *The New Palgrave Dictionary of Economics*, 2nd ed., ed. Steven N. Durlauf and Lawrence E. Blume (Palgrave Macmillan, 2008), Vol. 7, pp. 328–334.

39. A variety of authors have challenged the notion that space research stimulated the economy. A book of some fame in this respect was Amitai Etizioni's *The Moon-Doggle, Domestic and International Implications of the Space Race* (Garden City, 1964).

40. G. Genta and M. Rycroft, *Space, the Final Frontier?* p. 26. The amount in question is about ten times the budget for NASA.

41. Extrapolating from data published in www.spacefoundation.org/research/TSR07.php?id=418

42. Reported by Caron Alarab, "Gotta Have It," *Detroit Free Press*, August 25, 2005.

43. Satellites may prove crucial in monitoring the effect on the polar caps of the average global temperature rise.

44. Of course, the change in weather may also be beneficial to some areas. A warm Siberia, for example, may become one of the largest gardens of the world.

45. Peter E. Glaser, "Power from the Sun: Its Future," *Science* 162, no. 3856 (November 22, 1968): 857–861.

46. For details see Gerard O'Neill's *The High Frontier*, 2nd ed. (Anchor Press/Doubleday, 1982). See also T. Heppenheimer, *Colonies in Space* (Stackpole Books, 1977).

47. O'Neill, *The High Frontier*; Heppenheimer, *Colonies in Space*.

48. These are points where the gravitational pulls of the Earth and the Moon on a body balance with the centrifugal force—with a zero net force. A city placed in one of them would be in a stable orbit and would not require frequent corrections in its motion.

49. Harvey Feingold et al., "Space Solar Power: A Fresh Look at Generating Solar Power in Space for Use on Earth," Rpt, SAIC-97/1005, April 4, 1997.

50. Wendell Berry in *Space Colonies*, ed. Stewart Brand (Penguin Books, 1977), p. 36.

51. Berry, *Space Colonies*.

52. Berry, *Space Colonies*, p. 37.

53. Meadows, *Space Colonies*, p. 40.

54. And then, by all appearances, he piles abuse on top of bad argument.

55. Some question, for example, the belief that in just a few years we could build an entire ecosystem from scratch, as would be required in one of O'Neill's space colonies. In addition to that, proponents of the exploitation of the resources of the solar system are often very optimistic about doubtful technologies; for instance, they frequently make references to self-replicating machines. The implausibility of such machines, also called "von Neumann machines," will be discussed in Chapter 8.

56. Clark, *Space Colonies*, p. 38.

57. "Brave New World of Biosphere 2?" *Science News* 150, no. 20 (November 16, 1996): 312–313. The relationship with Columbia University ended in 2003.

58. This vision is well expressed in C. S. Cockell, *Space on Earth: Saving Our World by Seeking Others* (MacMillan, 2006).

59. James S. J. Schwartz, "Our Moral Obligation to Support Space Exploration," *Journal of Environmental Ethics* 33 (Spring 2011): 67–88.

60. The U.S. military is taking seriously the possibility of solar power satellites. And the Space Frontier Foundation sponsors a website for an open discussion of this technology: http://spacesolarpower.wordpress.com/

61. R. Winston, "Nonimaging Optics," *Scientific American* (March 1991): 76–81. See also Julio Chaves, *Introduction to Nonimaging Optics*, 2nd ed. (CRC Press, 2015).

62. The standard fusion reactor design uses a deuterium-tritium reaction, which produces neutrons, which in turn generate radioactive materials in the metal structure of the reactor. See Zubrin, *Entering Space*, particularly pp. 84–90 and 158–163

63. *Orbital Debris*, NASA CP-2360.

64. Until recently, NASA had a policy of bringing about technological breakthroughs with each new mission. Because of budgetary constraints, that policy apparently has changed. Many future missions may depend on a recycling of existing technology. If so, the scientific exploration of space need not drive *space* technology substantially anymore. The impact of the esoteric technology used to explore Jupiter and Saturn on the general technology cannot be discounted, but estimating that impact precisely, or even approximately, is not an easy matter, as previous remarks indicate.

3

The Philosophy of Exploration

Neither history, nor economics, nor the natural sciences seem to provide us with a solid argument for the exploration of space, but I will argue in this chapter that philosophy of science can. One of the main purposes of philosophy of science is to analyze the nature of science, and the main issue before us is whether there is something about the nature of science that creates the conditions for serendipity. For if serendipity is a natural consequence of science, then science will be practical in a very profound way, and we will have an answer to the concerns of the social critics. As I will argue, this philosophical answer will also allow us to meet the most crucial ideological objections.

The notion that science is deeply useful or practical captivated some of the early philosophical supporters of scientific investigation, men like Sir Francis Bacon in England (1620) and René Descartes in France (1637).

Descartes, whose development of analytic geometry had much to do with the eventual success of the Scientific Revolution, postponed the publication of his work on physics when the Catholic Church persecuted Galileo for defending views of the universe that disturbed the accepted harmony between man and God. Having to keep his research hidden, Descartes lamented, might be a grave sin against "the law that obliges us to procure the general good of mankind."[1] For as he saw it, "one might reach conclusions of great usefulness in life and discover a practical philosophy . . . which would show us the energy and action of fire, air, and stars, the heavens, and all other bodies in our environment, as distinctly as we know the various crafts of our artisans."[2] Once in possession of that knowledge, we may apply it, as we apply those crafts, "to all appropriate uses and thus make ourselves masters and owners of nature."[3]

Descartes's suggestive view, like Bacon's, fails to meet the challenge I have undertaken in this work: invoking it begs the question against the ideological critics, for it assumes what they most vehemently disagree with—that if we wish "to procure the good of mankind," we should practice science. Furthermore, it does not show us how scientific exploration and serendipity are related. It is clear, then, that we need a new argument.

The Dimming of Starlight. Gonzalo Munévar, Oxford University Press. © Oxford University Press 2023.
DOI: 10.1093/oso/9780197689912.003.0003

A Philosophical Case for the Serendipity of Science

My argument has two parts. The first part establishes a strong connection between scientific change and serendipity. The second part establishes a strong connection between scientific exploration and scientific change.

Argument Part 1—Scientific Change and Serendipity

1. *Scientific views are instruments for interacting with the universe, and they tell us what the universe is like.*

This statement may sound plausible, but it is by no means universally accepted. According to a still popular view, scientific knowledge is objective, objective is equated with factual, and as a result, facts become the business of science. From this, it presumably follows that the function of science is to collect facts about the universe, and that the function of space science is to collect them from space.

This conventional, and relatively static, picture of science makes it difficult to defend space science. If radiation from a certain region of the cosmos has been coming toward us for millions of years, and will be coming for millions more, what is the hurry to put a telescope in orbit to observe it right now?

The answer lies in a different notion of science: a notion that was first brilliantly developed by Galileo in 1632 and finally recovered by Paul Feyerabend and Thomas Kuhn in the twentieth century.[4] It is the notion that scientific views or theories are like spectacles through which we experience the universe.[5] And as Kuhn taught us, they are also instruments we use to interact with the universe.

To illustrate how far the shift on emphasis away from the collection of facts can take us, briefly consider the way in which Galileo surmounted one of the main obstacles to Copernicus's idea of the motion of the Earth.

In his analysis of the history of science, Paul Feyerabend explains how the Aristotelians employed the Tower Argument to show that the "facts" refuted the Copernican view. Suppose, with Copernicus, that the Earth moves. If you then drop a stone from the top of a tall tower, by the time the stone hits the ground the tower will have moved with the Earth, and thus the point of impact will be a considerable distance from the base of the tower. For the impact to be as close to the base of the tower as it actually is, the stone should follow a parabolic motion. But it obviously falls straight down (Figure 3.1). Thus, the supposition that the Earth moves cannot be correct.

We now know, however, that Copernicus was right—the Earth moves. But why should his view fly in the face of such obvious facts as the vertical, downward motion of the stone? As we see the stone leave the tower, we find it natural to say that the stone moves straight down. But this "natural interpretation" assumes, with Aristotle, that a normal observer under normal conditions can determine the motion of the stone. "Real" motion is presumably observable motion: a change in location that we can measure (in this case a normal observer functions as the measuring instrument: the motion of the stone is what the observer sees the stone do). In contemporary terms, the Aristotelian opponents of Copernicus and Galileo assumed that motion was operational (determinable by measurement).

Galileo's ploy, according to Feyerabend, was to offer a different set of "natural interpretations" according to which we do not actually observe the real motion of the stone. There are, Galileo explained, two components to the

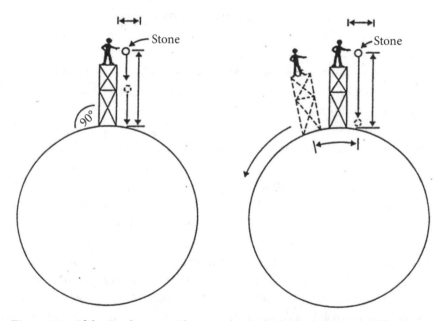

Figure 3.1. If the Earth moves, the stone lands far from the bottom of the tower. First published in my book, *A Theory of Wonder* (2021). (Reproduced courtesy of Vernon Press, Wilmington, DE)

motion of the stone: a straight motion toward the center of the Earth and a circular motion that it shares with the Earth ("circular inertia"). The stone, the tower, and the observer all share this circular inertia. But shared motion cannot be observed (when flying in a jetliner, we do not "see" the passenger in the next seat as cruising at 500 miles per hour, even though we may know that he is). Thus, we can perceive neither circular inertia nor the real (compound) motion of the stone. The normal observer sees only that component of the motion of the stone that he does not share: the motion toward the center of the Earth. Therefore, concludes Galileo, the stone does not fall straight down—it only seems to.

In this manner, Galileo defused one of the main objections to the Copernican view: the crucial "facts" his opponents adduced make theoretical assumptions. Certainly, Galileo points out, only if we already believe that the Earth does not move, we must conclude that the motion of the stone is "straight and perpendicular."[6] But if we believe instead that the Earth rotates, we must conclude that the "motion would be the compound of two motions," as well as parabolic (a "slanting movement," see Figure 3.2).

Galileo's logic leads to a straightforward conclusion: *what the "facts" are depends on what scientific theory we already accept.* Clearly, Galileo says, Aristotle, Ptolemy, and their followers "take as known that which is intended to be proved."[7]

Galileo's opponents, once again, had an operational concept of motion: motion is observable change of position over time. Galileo proposed instead a concept of compound motion in which one of the components is theoretical (and, in principle, unobservable, since the observer shares it).[8] He thus defused the objection, not by advancing a set of facts *free* of theoretical assumptions, but a set of facts with *different* assumptions. The new "real" motions of objects were no longer directly observable, and the relativistic basis for holding this view introduced a new way of doing physics.

Different views of the universe thus lead to different assumptions, and different assumptions lead to different evaluations of what is to count as evidence, as facts. They also lead, as in the case of the Copernican Revolution, to a profound and radical transformation of our understanding of what the world is like.

In the nineteenth century, William Prout suggested that all elements are formed from hydrogen. He thus predicted that all atomic weights should be multiples of 1, since that is the atomic weight of hydrogen. But careful measurements of the atomic weights of several pure elements, chlorine,

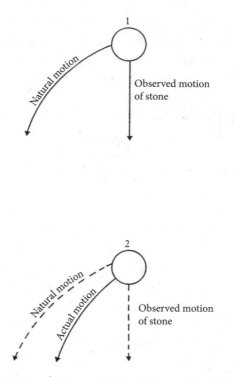

Figure 3.2. The new natural interpretation of the motion of the stone. The natural motion is given by circular inertia, which the stone shares with the Earth and the tower. First published in *A Theory of Wonder* (2021). (Reproduced courtesy of Vernon Press, Wilmington, DE)

for example, did not produce whole numbers. As in the case of the Tower Argument and similar experiments,[9] the seemingly impregnable facts refuted an interesting new view.

A century later, though, a new atomic theory explained that each element differs from others by the number of protons in the atomic nucleus (exactly matched by the number of electrons "around" it). The new theory, however, also proposed particles called "neutrons" in addition to protons and electrons. Neutrons, which have mass but no charge, are also located in the nucleus, but their numbers can vary. That is, there can be several varieties of the same element (isotopes), *with different atomic weights.*[10] The extremely careful nineteenth-century measurements of the atomic weight of pure chlorine turned out to be measurements of the mix of chlorine isotopes found in nature. The atomic weight of each of the isotopes is indeed a whole number, much in accordance with Prout's insight. A revolutionary theory, once more, changed in this case not only the "facts" but also our understanding of what an element is.[11]

According to Kuhn, scientific views (which he called "paradigms") tell us what elements there are in the world and what relations exist between these elements. Suppose, for example, that scientists come to view the basic components of the world as little "billiard" balls in frequent collisions. As soon as we begin to view the world in this manner, a host of new problems requires solutions. If a collision is perfectly elastic, what happens to the particles? (The system's momentum going in must equal its momentum going out.) What happens in inelastic collisions? (Some energy will be dissipated, probably as heat.) Reasonably precise solutions to such problems are found using the laws and mathematics of a new science (e.g., Newton's), which are tailored to the kinds of problems that arise because we think the world is guided by the analogy to little billiard balls.

How we view the world thus determines what sorts of problems are meaningful and what kinds of solutions are acceptable. This is, of course, a rough and tentative way of describing science, but it suggests the sense in which we can speak of science as a pair of spectacles that permits us to see the world. I am speaking metaphorically, to be sure. But that metaphor is by no means farfetched. We should realize that without our scientific views we would simply be blind to many aspects of the universe.

Sight itself is more than a mechanism for forming mental images of the world: it is also a complex means of interaction with the world. Consider two examples. As you walk down a familiar street at night, you vaguely make out in front of you some amorphous shapes. Suddenly, you hear the nasty growl of an attack dog, and almost instantly one of the amorphous shapes turns into the well-defined—and frightening—body of a Doberman. The threat we hear informs and changes what we see.

In a famous experiment, F. P. Kilpatrick invited subjects to look at two rooms through peepholes.[12] They first looked at a normal (rectilinear) room. The next room was distorted in that the left wall was twice as long as the right, but the subjects noticed nothing unusual about it. They were then asked to touch a mark on the back wall with a long stick. Since they saw the room as normal, they also saw the back wall as parallel to the front one, and so their attempts to touch the mark failed. Experimenting with the stick led them eventually to success, and as soon as they achieved it, they immediately saw the room as distorted.

These two cases illustrate how the visual cortex circuits that let us see the "properties" of objects (what people understand by visual perception) are meant for flexible interaction with our environment. Another, faster circuit allows us to react almost instantly without even making "pictures" of the object. A professional tennis player, for instance, will react to the flight of the ball even before consciously seeing the ball (so-called muscle memory).[13] And animals that live in completely different environments emphasize completely different senses and the brain structures that support them (bats use sonar; some fish use electric fields).[14] In general, animals perceive as they do because their senses were of practical advantage to their ancestors as they interacted with the world.

Scientific views also give us more than pictures or representations of the universe. Like magnified senses, they provide means of interacting with the universe, for they ask questions by seeing, probing, and touching nature at many different energies and magnitudes. Thus, when we learn to "see" the universe, we actually learn to make contact and deal with its diverse facets in many different ways.[15]

This interactive view of science is controversial among philosophers, and it may sound strange to scientists who believe that the facts always decide the worth of scientific ideas—never the other way around. But it receives strong support from the history and the practice of science (especially at crucial scientific junctures, such as Galileo's defusing of the Tower Argument, where the facts we choose depend on the theories we favor), as I have argued elsewhere.[16] The point of this excursion into a central controversy in the philosophy of science is that scientific exploration in general, and space exploration in particular, creates circumstances that force science into such crucial junctures. Exploration, therefore, as we will see later, makes inevitable the radical transformation of the ways we "see" and interact with the universe. And from that, serendipity is practically inevitable.

I place quotation marks around "see" to prevent a pointless but surprisingly common misinterpretation of the approach under discussion. Some philosophers take this interactionism to imply that with changes of worldview come changes in the actual visual perception of the world. Sometimes this may happen: expectation does affect perception, and different theories may set up different expectations. But it need not happen in the cases when we are trying to decide between two theories. Galileo saw the stone fall straight down, just as the Aristotelians did. And Prout and his critics all saw the same numbers on their instruments. Same perceptions but different facts. Galileo's argument destroyed not only the empiricist distinction between theory and fact but also the empiricist connection between perception and fact.

Nevertheless, this is not a book about the limitations of empiricist theories of science. My intent is rather to apply and illustrate the findings of the inter-actionist approach to the nature of science. I trust that those applications and illustrations will best show the worth of such findings.[17] In any event, with those findings in mind, our problem of justifying space exploration will take on an entirely new light as we place emphasis on the essential transformations of science and their consequences.

    2. *Scientific views determine what problems, dangers, and opportunities we can be aware of.*

Since our worldviews tell us what the world is like, they also determine ultimately what opportunities we can take advantage of, and what problems and dangers we can be warned about. And thus, it follows that:

    3. *With changes of worldview come the realization of new problems, dangers, and opportunities.*

As we have seen earlier (Point 1), those changes can be profound, which means that as our science is radically transformed, so is our panorama of problems, dangers, and opportunities.

    4. *By becoming aware of new problems, dangers, and opportunities, we also become able to think of new solutions and new technologies.*

Einstein began his career by asking "useless" theoretical questions such as "What would the universe look like if I were traveling on a ray of light?" In trying to satisfy his curiosity about this and other equally "impractical" is-sues, he was led eventually to develop his theory of relativity and to take a decisive role in pushing physics toward quantum theory with his work on the photoelectric effect (although he later disagreed with the full-blown quantum physics of Bohr and Heisenberg). In these and other respects, he changed several of our views of the world in profound ways, opening in the process the opportunity for a new understanding of light. This new understanding led to the theory of lasers. Lasers, in turn, opened up many technological opportunities. It was not long before some researcher decided to apply them in medicine.[18] Today lasers are used in extremely delicate surgical procedures that would not be possible with any other technology known previously to medical practitioners. And it all began with a change of worldview in a field far removed from medicine at the time.

In contrast, imagine a crash program in Einstein's early days to have surgeons develop a surgical instrument that could do the sorts of things that a laser can do today. Is it reasonable to suppose that the point would have come when the well-funded surgeons would have realized that their aim required the overthrow of the physics of their day? And would they have then laid out the steps necessary to replace that physics with a view that would lead them to lasers, and so on? I think not. But without the new physics, could the surgeons have developed the equivalent of lasers, or of much other Western medical technology for that matter? Again, I think not.

These considerations undermine the objection that all the good indirect results (e.g., the spinoffs) of the space program can be achieved by spending the money directly in the relevant areas: for benefits in one area may well require a prior radical transformation in another.

Such theoretical transformations make us aware of possible new solutions and opportunities, and that is precisely what we mean by serendipity.

Having established that:

1. Scientific views are instruments for interacting with the universe, and they tell us what the universe is like.
2. Scientific views determine what problems, dangers, and opportunities we can be aware of.
3. With changes of worldview come the realization of new problems, dangers, and opportunities.
4. By becoming aware of new problems, dangers, and opportunities, we also become able to think of new solutions and new technologies.

We may then conclude that:

5. *Serendipity is the natural (practically inevitable) result of scientific change.*

Argument Part 2—Scientific Exploration Leads to Scientific Change

Changes in worldviews are inevitable, given the nature of science. The reason is that worldviews (e.g., comprehensive scientific theories) are our creations and thus imperfect.[19] Therefore, they are always in need of refinement, modification, or replacement.

The pressure for such changes comes from the exposure both to unusual circumstances (which force us to stretch our views) and to competing ideas (which are often developed to account for a few of those unusual circumstances, and then extended to explain the entire field). And—here is a key point—scientific exploration, by its very nature, places science in new circumstances and presents it with new ideas. Thus, scientific exploration leads not merely to the addition of a few, or even many, interesting facts but to the transformation, perhaps the radical transformation, of our views of the world.

Let me rephrase this second part of the argument in schematic form:

6. *Science is dynamic.* Science is always changing because
   (a) *It is never complete* (being a human creation).
   (b) *When challenged by new circumstances, it must adapt (i.e., change).* For example, cancer and HIV/AIDS have challenged scientists to alter profoundly our ideas about cell functioning and development. And astronomy has undergone many radical changes motivated, in great part, by new instruments that have allowed us to look at hitherto unimagined aspects of the universe (e.g., Galileo's telescope and the discovery of the phases of Venus and the moons of Jupiter, as well as the recent upheaval created by the discoveries made with the new generations of telescopes).
   (c) *When challenged by new ideas, science is, once again, spurred to change.* Think of the radical transformation of biology as the result of Darwin's idea of natural selection.
7. *Scientific exploration places science in new circumstances and presents it with new ideas.*

This is true almost by definition. When we explore scientifically, we either move science into new areas or else think about it in a new way (i.e., in the light of new ideas). I say "almost" for two reasons. First, even though I am presenting a conceptual argument, I do not wish to engage in a semantic dispute. Perhaps someone might give an example in which a scientist explores without placing science in new circumstances or thinking about science anew. Nevertheless, these two activities cover the range of what scientific exploration characteristically does (in a strong sense of "characteristic"). This is why I aim for the conclusion that serendipity is a *natural* consequence of scientific exploration, where "natural consequence" means a (strongly) characteristic or practically inevitable consequence. A practically inevitable consequence of having a human genome is to be born with one head, one heart, two eyes, and two legs. But some humans are born with only one leg, say, and some human embryos do not even get to be born.

The second reason is that some very scholarly critics may feel that "exploration" is too romantic a name for what scientists do. Scientists presumably ponder, observe, investigate, and carry out experiments, but they are explorers only in a metaphorical sense. I do not wish to engage in a semantic dispute over this issue either, although it would be peculiar that so many people from so many walks of life should understand perfectly what I mean when I talk about the scientific exploration of space, and that they should themselves talk this way, if "exploration" is indeed the wrong term. In any event, scientists' attempts to satisfy their curiosity about the universe do lead them into new areas and do motivate them to look at their collective understanding in new ways. This is all my argument needs.

Moreover, these two activities provide the natural conditions for change in science. Given that science is not complete, when it is placed in new circumstances (e.g., in dealing with significantly new phenomena, or being applied well beyond its domain), it is characteristically forced to adapt (change). The challenge of new scientific ideas is an additional factor in bringing about scientific change.[20]

From Points 6 and 7 we may conclude, therefore, that:

8. *Scientific exploration leads to change in our scientific views.*

As I argued earlier, the crucial feature of science is not merely the addition of a few, or even many, interesting facts but the transformation, perhaps radical, of our views of the world (cf. my remarks on Feyerabend, Galileo, and the Tower Argument).[21] This essential feature turns serendipity into a natural consequence of a dynamic science. If science is to be dynamic, it must be challenged, and it must change. But the change that matters is the transformation of our views of the universe. For once we think about the universe differently—once we have a different "communal" perception of it—we come to perceive also hitherto unknown dangers, new solutions, and new opportunities. Such is the cradle of serendipity.

Putting the conclusions of Parts 1 and 2 together, we arrive at the conclusion of the whole argument:

9. *Since scientific exploration leads to change and scientific change leads to serendipity, exploration leads to serendipity.*

From Serendipity to Justification

Clearly, then, a dynamic science makes certain things possible that otherwise would be not only beyond our reach but also beyond our imagination. That is one of the main reasons why we cannot afford not to do science: many problems we face, knowingly or not, cannot be solved unless a different point of view comes into play. Likewise, many opportunities would never come our way. Therefore, to reject or slow down the process by which science grows, by which we refine and replace our communal spectacles, amounts to a decision to deprive ourselves of much that is good and to continue to expose ourselves unnecessarily to who knows what dangers.[22]

Supporters of space exploration can now justify their expectation of a bounty from space precisely because exploration presents many challenges to our science and technology. Since space exploration is thus so likely to contribute to the transformation of our views, investing in it has a clear advantage over investing in fields not so ripe with challenge.

Knowing that serendipity is a natural consequence of science, the supporter of exploration may now say with Descartes that a failure to explore is a failure to carry out our obligation to "procure the general good of mankind." The justification the supporter can now offer for exploration in general, and for the heart of space exploration in particular,[23] sounds like a practical case, albeit more subtle and indirect than the one made in Chapter 2. But it is a practical case born out of the nature of space science itself, and thus the guarantees that it offers go well beyond those of historical anecdotes.

This deeper and fundamental practicality forms the basis of the supporter's response to the strongest social objections. The supporter of exploration can now explain to the social critic why the previous benefits of exploration were not a fortunate accident: they were the result of the inevitable expansion of opportunity that is part and parcel of scientific exploration. Once we understand the dynamic nature of science, we are in a position to vouch for its future serendipity. As we saw in the previous chapter, to a casual observer the heart of space exploration may not appear to have obvious practical benefits. But this deeper investigation reveals a long-term, fundamental practicality: the practicality that comes when a transformation of our views of the universe expands our range of opportunity.

As the benefits from exploration become routine, the frontier of the unknown is pushed further out into the cosmos and our challenges shift accordingly. In indulging our enthusiasm, we are thus bound to force a change in our panorama of problems and opportunities.

The argument works against the ideological objection as well. Unless we commit species suicide, we will continue to interact with the Earth and transform it in small and large ways. By doing so, we act in the manner of all other living beings: a tree grows tall and gives shade to violets and mushrooms that could not have lived without it; a beaver builds a dam, harming rose bushes and fish, but helping water lilies and frogs; and once upon a time, bacteria gave the Earth its oxygen and nitrogen atmosphere.

The question for us is not whether we will interfere, but rather how much and how wisely. Now, to act wisely we need knowledge of ourselves, of the Earth, and of our interactions with the Earth. Otherwise, we are likely to impose too big a burden on the planet or on its human inhabitants.

Such knowledge is not complete as of this writing—not even environmental activists can reasonably claim that they know everything about our planet—and it may never be complete. That is, our perspective is limited, and therefore we need a dynamic science that can change our panorama of problems and opportunities. To eschew dynamic science is to deprive humankind of the chance to act wisely. We are already a big presence on the Earth and need to move carefully in the dark of our ignorance. It would thus be irresponsible to forgo the lanterns that may illuminate our way (lanterns such as NASA's Mission to Earth, to be discussed in the next chapter). Space exploration is thus not a false panacea but an important means to a cleaner and better future.[24]

Challenges to the Argument

In a survey the journal *Science* did in 1964, only 16% of the science PhDs who responded agreed with President Kennedy's decision to go to the Moon, while an overwhelming 64% disagreed. It was generally felt then that the Apollo program was undertaken mainly for political reasons.[25] Ever since, many scientific and other critics have questioned the scientific value of space exploration.

The point is that my serendipity argument depends on a close connection between space exploration and science. If this connection is brought into question, my argument is also brought into question.

According to a second objection, it is not enough to show that the scientific exploration of space is serendipitous. We are still required to show that such exploration is likely to produce greater serendipity than competing activities, including other types of scientific exploration.

Let me describe these objections in greater detail.

1. *Space exploration does not involve fundamental science in a significant way.*

The first objection goes something like this: what my argument shows, strictly speaking, is that changes in *fundamental* science lead to a different panorama of problems, solutions, and opportunities, hence the serendipity of science. Such changes have the desired effect because fundamental science gives us a way of viewing the universe and of interacting with it.

It is not clear, however, that applied or peripheral science would have similar effects. And to these critics it seems that the science done in the pursuit of space exploration is, for the most part, applied or peripheral.[26] This is not to say that space exploration is not likely to produce serendipity of some sort, for obviously it already has. The point is rather that the significance of what space exploration will accomplish is much less than I have made it out to be. Yes, we will have some interesting though marginal science and lots of gadgets, but no radical transformations of our main points of view.

Two considerations may tempt us to dismiss this objection summarily. The first is that it assumes too regal a status for fundamental or pure science as compared to applied science and technology. A moment's thought makes us realize that "gadgets" have often driven revolutionary developments in fundamental or pure science: lasers are pivotal instruments in the study of fusion; personal computers have enabled the launching of hitherto undreamed-of theoretical work in many scientific disciplines, from mathematical physics to neuroscience; and let us not forget the most famous influential gadget in the history of science, Galileo's telescope, which was invented as a toy in Holland.

It is clear, then, that transformations in technology and "applied science" can create a new panorama of problems and opportunities for the practice of fundamental science.

Nevertheless, I will not take the easy way out offered by this consideration. The science done in space exploration runs the gamut from the most applied to the most fundamental, as I hope to show in the rest of the book, and thus it brings out the deep practicality of science in its fullest sense.

The second consideration is that the critic who belittles the scientific value of space exploration is perhaps a bit of a straw man: space science is far more respectable now than in the days of President Kennedy. Three reasons, however, should keep us from deriving much reassurance from this consideration.

The first reason is that not all space science is now respectable, as we will see shortly. In any event, it is important to understand why that shift of perception took place in the fields of space science where it did.[27] The second reason is the need to address the nagging suspicion that some space research has gained prominence purely because the government has thrown big money to support it. If this suspicion is correct, society's quest for space exploration has distorted the practice of science. The third reason is that even if, contrary to fact, most scientists did have a high opinion of most space science today, it would still be useful to state as bluntly as possible why someone might not agree. For in replying we stand to explain better *why* we ought to go into space.

Space science covers many fields, but for the purpose of this book they can be subsumed under three main categories: planetary science, space physics and astronomy, and space biology. Let us see why their serendipity might seem questionable.

Take planetary science (under this rubric I am including comparative planetology and the scientific exploration of the solar system in general).[28] Granted that by going into Earth's orbit and looking down we can learn much about our own planet; but what can we learn about the Earth from looking at another planet? It would seem, as an early Greek might say, that if the other planet is different, we are not learning about the Earth; and if it is like the Earth, we should not waste effort going there when we might as well look at the Earth itself.[29]

As for space physics and space astronomy, how can they change our lives down here? It may be fascinating to find out what makes quasars burn; but fascination aside, will that knowledge feed hungry children or at least make automobiles run more efficiently? We need to see how space physics and astronomy can come to be in a position similar to that of the revolution in physics that led to the laser and its use in medicine.

A critic might argue that lasers are built on fundamental principles of matter; on principles, furthermore, that apply right here on Earth. So there is no mystery why a revolution that gave us those principles had terrestrial applications. By going into deep space (see Figure 3.3), by placing telescopes in orbit and all that, we might challenge our points of view and force them to change. But they are points of view about what is up there, not about what is down here. Or are they?

Space biology fares even worse, for many space scientists themselves see little value in it beyond the need to keep astronauts healthy. And since many of those scientists would prefer uncrewed exploration, even this conditional value of space biology is in question. Such is a common verdict regarding

Figure 3.3. The Andromeda Galaxy (M31) is our sister galaxy in our local galactic group. Located approximately two million light years away, M31 is a spiral galaxy believed to house a supermassive black hole at its center. Although similar in size to the Milky Way, M31 is twice as massive, containing approximately one trillion stars. Many areas of nebulous star formation are visible in the image. This image is about twenty hours' worth of data shot with a Celestron RASA 8 (see color plate). (Image courtesy of Michael R. Shapiro)

the branch of space biology that investigates the behavior of *terrestrial* life in outer space. Another branch of space biology, astrobiology, investigates *extraterrestrial* life. Under this rather cryptic description, astrobiology became a target for critics who derided it, until the Mars meteorite controversy in the 1990s, on the grounds that, since we have never found extraterrestrial life, astrobiology investigates nothing at all.[30]

I will provide replies in the next three chapters, one per field. It will become evident, however, that this division of space science is largely a matter of convenience, for strong connections exist among the three fields. Indeed, the seeds for the answers to some of the questions pertaining to astronomy and biology will be planted in the discussion of planetary science.

> 2. *The serendipity of space exploration may not be greater than that of other scientific enterprises.*

If we are to support scientific exploration because its serendipity will reward us with the tools to improve life on Earth, are there no better candidates than space exploration? Consider oceanography, for example.[31] It is clear that the oceans play a crucial role in our climate and in the planet's ability to sustain life. The benefits of understanding the oceans better thus seem quite direct. Shouldn't oceanography then have greater priority than space exploration?

I offer two replies. The first is that I have never argued that space is the only stage for scientific exploration. In a well-run world, space exploration would be one of the important tasks human beings undertake, and perhaps some other scientific tasks should have even greater priority.

The second reply is that the priority of space is likely to be very high anyway. Consider the example of oceanography again. Clearly, obtaining knowledge of the oceans is very important to us. But as we will see in the following chapter, success in securing that sort of knowledge will require, at least in part, a global approach to the study of the oceans and the other systems with which they interact—a global approach for which space technology is exceptionally well suited. My suggestion is, then, that the majority of serious "competitors" to space exploration will actually be more successful if done *in conjunction* with space exploration.

Of course, I do not wish to claim that all space exploration is scientific. As space activities become routine, more and more of them turn into industrial enterprises or financial investments (e.g., satellite communications). The aim of this chapter was to provide a philosophical case, via serendipity, to justify the heart of space exploration.

In overcoming the objections, supporters of space exploration will be able to appropriate Descartes's words when claiming, for example, that space biology will contribute to medicine and thus bring about "the preservation of health, which is without doubt the chief blessing and the foundation of all other blessings in this life."[32] And in addition they may proudly look forward to the new mastery of nature with which space science will reward their efforts. For that mastery will lead to "the invention of an infinity of arts and crafts which [will] enable us to enjoy without any trouble the fruits of the earth and all the good things which are to be found there."[33] To the fruits of the earth, the supporters will say, space exploration promises to add the bounty of the universe.

Notes

1. Rene Descartes, *Discourse on Method*, in *Descartes's Philosophical Writings*, trans. and ed. E. Anscombe and P. T. Geach (Thomas Nelson and Sons, 1969), p. 46. For some passages from the *Discourse* quoted later, I will favor the translation by H. S. Haldane and G. R. T. Ross in *The Philosophical Works of Descartes*, Vol. I (Cambridge University Press, 1972).
2. Descartes, *Discourse on Method*.
3. Descartes, *Discourse on Method*.
4. Galileo, *Dialogues Concerning the Two Chief World Systems*.
5. The most important references in this regard are Paul K. Feyerabend, *Against Method* (NLB, 1975); Thomas S. Kuhn, *The Structure of Scientific Revolutions*, 2nd ed. (University of Chicago Press, 1970); Imre Lakatos, *The Methodology of Scientific Research Programmes* (Cambridge University Press, 1978); Karl R. Popper, *The Logic of Scientific Discovery* (Hutchinson, 1959).
6. This and the following quotations from Galileo come from his *Dialogue Concerning the Two Chief World Systems* (Modern Library of Science, 2001), p. 162. The original was published in 1632. On p. 198, Galileo restates the argument. One ought to say, he claims, the following: If the earth is fixed, the rock leaves from rest and descends vertically; but if the earth moves, the stone, being likewise moved with equal velocity, leaves not from rest but from a state of motion equal to that of the earth. With this it mixes its supervening downward motion, and compounds out of them a slanting movement.
7. Galileo, *Dialogue Concerning the Two Chief World Systems*.
8. From a spaceship today, we could see the parabolic motions of falling objects, but this option was neither practically nor rhetorically available to Galileo. And, at any rate,

as long as the observer shares the rotation of the Earth, the real motion of the falling rocks would remain in principle unobservable to him.

9. Galileo proposed many experiments, some with cannonballs, to arrive at the same conclusion of the Tower Argument: that the Earth could not move. That is, he made the opponents' position even stronger than his actual opponents had, and then demonstrated that such a position was question begging. Galileo, "Second Day."

10. Prout proposed his idea in 1815. By the early 1900s, Hantaro Nagaoka had proposed a "Saturn" model of the atom and Lord Kelvin and J. J. Thompson had defended the "plum-pudding" model. It was not until 1911, though, that Ernest Rutherford's idea of an atomic nucleus that contained both protons and electrons could perhaps begin to support Prout's hypothesis. The discovery of the neutron, the full understanding of isotopes, and the full retrospective vindication of Prout would have to wait until James Chadwick's experimental results in 1932. A very readable account appears in D. Lincoln's *Understanding the Universe: From Quarks to the Cosmos* (World Scientific Publishing, 2004).

11. Lakatos offers a particularly interesting philosophical account of this and similar incidents. Lakotos, *The Methodology of Scientific Research Programmes*.

12. Kilpatrick used three rooms, but I am describing only part of the experiment. *Explorations in Transactional Psychology* (New York University Press, 1961).

13. S. Blackmore, *Consciousness* (Oxford University Press, 2004), pp. 38–39.

14. G. Munévar, *A Theory of Wonder: Evolution, Brain, and the Radical Nature of Science* (Vernon Press, 2021), pp. 154–155.

15. For a more detailed account, see my *Evolution and the Naked Truth* (Ashgate, 1998). For a converging view from the psychology of perception, see Victor S. Johnston's *Why We Feel: The Science of Human Emotions* (Perseus Books, 1999).

16. In addition to the works mentioned in the notes above, including of course Galileo's, I would suggest reading two of my papers that summarize many of the main arguments: "A *Réhabilitation* of Paul Feyerabend," which can be found in *The Worst Enemy of Science? Essays in Memory of Paul Feyerabend*, which I edited with John Preston and David Lamb (Oxford University Press, 2000), pp. 58–79, and "Conquering Feyerabend's Conquest of Abundance," *Philosophy of Science* 69 (September 2002): 519–535. See also my new book, *A Theory of Wonder* (Vernon Press, 2021).

17. This situation mirrors the process by which Galileo's view, the Copernican theory, came to be accepted. Such a process is best explained within an evolutionary and neurobiological perspective of the nature of science. For an account of that perspective, see my *Evolution and the Naked Truth*.

18. I do not mean to suggest that this process is automatic or easy. In the case of the laser, it took a maverick with a combined background in physics and engineering, Charles H. Townes, to see the possibilities. It also took a great amount of persistence on his part, in the face of profound skepticism by the profession. Charles H. Townes, "Resistance to Change and New Ideas in Physics: A Personal Perspective," in *Prematurity in Scientific Discovery*, ed. E. B. Hook (California University Press, 2002), pp. 46–58.

19. Scientists like to draw a distinction between their use of the word "theory" in science and the ordinary use of the word. In science, the word is almost an honorific title given to a comprehensive set of ideas that at least begins to explain a range of phenomena; ordinary parlance betrays an empiricist bias (theory as less important than fact) and the word is almost derogatory, as in "evolution is just a theory," meaning "little more than a guess." This equivocation irritates scientists, for then quantum physics and general relativity would also be "just theories." To reduce the confusion, I keep pointing out that I am talking about comprehensive theories all along.

20. To put the point in a language closer to that of professional philosophers: by the "natural conditions" of change in science, I mean that (disjointly) they are *practically* necessary and sufficient. That is, first, one or the other is normally required for scientific change to occur (overcoming the natural inertia against intellectual and experimental retooling). And second, the challenge of new circumstances characteristically forces science to change (this is a practical, not a logical certainty, for the scientific field may instead fall apart, or society may stop funding research, or a supernova may destroy the world, etc.). This is not to say that science is bound to find solutions to its problems. There may be no cure for AIDS, for example, but in looking for it, researchers have profoundly transformed scientific medicine. The challenge of new ideas is often sufficient to bring about change as well (when those ideas offer significant alternative ways of looking at difficult problems, etc.).

21. This theory of science is developed in detail in my *Radical Knowledge: A Philosophical Inquiry into the Nature and Limits of Science* (Hackett, 1981).

22. We may also expose ourselves to new risks. I will discuss this matter in the last two chapters.

23. Cf. the remarks made at the end of Chapter 2 about those aspects of exploration that fire the imagination and motivate the conquest of the cosmos.

24. Historians may point out that the argument sketched in the two previous sections of this chapter is largely based on history and so they may wonder about my remarks in Chapter 2 against a historical case for serendipity. The answer is that here I place the history of science in a philosophical context. Several of my premises do require the history of science for their support, but my conclusions are not the results of inductive inferences from history. They depend instead on conceptual inferences about the nature of science and exploration. This is what makes my argument philosophical.

25. Even some supporters of exploration agree. Ben Bova from the National Space Society writes in a letter to *Science* (233, August 8, 1986, p. 610) that "The U.S. space program's primary motivations are, and always have been, political and economic." He also thinks that it is a myth "that the space program exists mainly for the purpose of scientific research." I will have more to say on these views in Chapter 7.

26. For an account of this attitude against space science, see my "Pecking Orders and the Rhetoric of Science," *Explorations in Knowledge* III, no. 2 (1986), pp. 43–48, reprinted in my *Evolution and the Naked Truth*, pp. 181–188.

27. This attitude within science, as well as negative attitudes about science in the larger society, plays an important role in our evaluation of space policy and of specific

proposals for funding space undertakings. This role is, however, seldom made explicit. We often have little more than a gut feeling about how priorities should be allocated. But would not a different idea of the nature of science—and of the nature of space science—influence our gut feeling?

28. For an account of the low status the planetary sciences suffered until rather recently, see Stephen G. Brush, "Planetary Science: From Underground to Underdog," *Scientia* 113 (1978), p. 771. Brush demonstrates how the prejudice against planetary science was blind to the history of physics.

29. This might be the approach taken by a student of Xenophanes.

30. This popular opinion of the field has changed considerably since David McKay's team's analysis of a now famous Martian meteorite (ALH84001) suggested that there were traces of fossil life inside of it. NASA has capitalized on the public enthusiasm, even though most meteorite experts have been hostile to the hypothesis. This issue will be discussed in Chapter 6.

31. This point was suggested to me by Terry Parsons.

32. Descartes, *Discourse on Method*, trans. Haldane and Ross, p. 120.

33. Descartes, *Discourse on Method*, trans. Haldane and Ross, p. 119.

4
Comparative Planetology and Serendipity

Science fiction gave us forests on the back side of the Moon, Martian canals constructed by advanced civilizations, and, in Venus, a throwback to happy early times: paradise. Unfortunately, the Moon is lifeless, Mars is a desert, and Venus is hell. As our knowledge of the solar system has advanced, we have moved our imagination beyond its confines. The worlds of strange intelligent creatures and monstrous beasts, of great wisdom or unparalleled horror might well exist—but around some distant star, safe from the rocket probes that might render empty what fiction has filled with the riches of dreams.

A social critic may wish to know why we should then want to explore the inhospitable worlds within our rockets' reach. Can there be, for example, any link between the exploration of Venus's poisonous atmosphere and the well-being of those who breathe our own atmosphere?

There is. There are many, in fact. Let me begin with one striking and important example of the serendipity of comparative planetology: the discovery of the threat to the ozone layer.

Ozone forms when oxygen molecules (O_2) capture oxygen atoms (O) to combine into larger molecules, ozone (O_3). Ozone acts as a nasty pollutant on the surface, particularly in the air of our large cities, but at high altitudes a layer of ozone absorbs ultraviolet radiation, thus protecting plants and animals on the surface from serious damage to their DNA. Indeed, life was confined to the oceans for much of the history of our planet, until the level of atmospheric oxygen increased enough to form a substantial ozone layer.

Now to Venus. When NASA scientists found fluorine and chlorine compounds in the atmosphere of Venus, they investigated the chemistry of those molecules and determined the rate constants of their chemical reactions. Those rate constants were later used by Sherwood Roland and

The Dimming of Starlight. Gonzalo Munévar, Oxford University Press. © Oxford University Press 2023.
DOI: 10.1093/oso/9780197689912.003.0004

Mario Molina to discover that chlorofluorocarbons (CFCs) destroy ozone in the presence of high ultraviolet radiation. That is, they discovered that the Earth's ozone layer might be in trouble. This discovery came as a shock to many researchers and industrialists, for CFCs had been developed precisely because, being "inert," they could not react with anything or harm anything. They seemed just perfect for use in air conditioners, refrigerators, and aerosol deodorant cans.

Unfortunately, high in the atmosphere, ultraviolet radiation breaks up the CFC molecules, and the freed chlorine atoms interact with the ozone, destroying it. This discovery was confirmed by Michael McElroy, whose group had the required tools because, as Carl Sagan pointed out, they were working on the chlorine and fluorine chemistry of the atmosphere of Venus.[1]

The presence of a large hole in the ozone layer over Antarctica was further confirmed by satellite data and later tracked and made vivid and dramatic by satellite pictures. This prompted scientists, industrialists, and governments, acting in concert, to ban CFCs, so as to significantly reduce the threat, which has been achieved (although it may take some forty years longer for the CFCs still in the atmosphere to dissipate and the ozone layer over Antarctica to fully recover).

This example is a beautiful illustration of the serendipity of comparative planetology. By investigating the atmosphere of Venus, we transformed our knowledge of planetary atmospheres; this knowledge made us aware of a serious problem here on Earth; and space technology helped us monitor the problem and provided the information needed to achieve a solution. And eventually we put the solution to the problem into effect.

Such is the link we seek between planetary science and the well-being of humankind: we need to explore the solar system in order to improve our views about the *Earth*. And we need to improve those views so that we may deal more wisely with certain social and environmental problems that could become acute in a few decades or outright disasters in the long run.

This example of the serendipity of exploring the solar system paying off here on Earth illustrates the following argument. To have a good grasp of global problems and their possibly serious consequences, we need to understand our global environment. But to understand the global environment of the Earth, it is important to understand the Earth as a *planet*. To understand the Earth as a planet, however, it is necessary to study the other members of the solar system. And, of course, to study the solar system well, we need to go into space.

Understanding the Earth as a Planet

A Sketch of the View

Our global understanding of the Earth depends on what we think a *planet* like Earth is like. The mechanisms that regulate the Earth's environment interact with each other in many loops and cycles run by energy, and that energy comes either from the Earth itself or from extraterrestrial sources. The energy that comes from the Earth depends very much on the sort of planet the Earth is. And the useful energy that comes from the Sun and the rest of the solar system depends on how the *planet* Earth relates to the rest of the system.

The Earth produces heat that rejuvenates its surface by creating, moving, and breaking up continents; by forming mountain ranges when the tectonic plates that carry the continents collide; by bringing new materials from the mantle into the crust, oceans, and atmosphere through volcanoes and mid-ocean ridges; and by recycling lands and gases through the spreading and subduction of the crust (the sinking of one plate under another when they collide). The rejuvenation of the Earth's surface creates a great variety in the environment, one of the crucial factors in the natural selection of living things. Moreover, the energy injected into the oceans and the atmosphere drives those systems and greatly influences how they interact with one another.

The Earth's gravitational energy keeps the atmosphere from dissipating into space, and thus it determines to a high degree the density of that atmosphere. That density, in turn, influences the chemistry of the environment and the climate of the planet. To see the point clearly, it pays to compare our planet with others. For example, the density of Mars's atmosphere is so low (about one-hundredth that of Earth's) that water cannot exist in liquid form: it goes directly from ice to vapor.

Internal heat and gravitation are both functions of the mass and structure of the Earth. Let us discuss heat. The main sources of internal heat are the release of energy from the decay of radioactive elements such as thorium, uranium, and potassium in the interior of the planet, and the energy left over from the gravitational collapse of the matter that formed the planet initially. Associated with this second source is the heat from differentiation, which occurs when denser materials move downward and displace less dense materials toward the surface. And initially there was also, of course, the extraordinary heat that the bombardment of the Earth by asteroids generated, enough, in the bigger collisions, not only to vaporize oceans and atmosphere but to melt the entire surface.

Our ideas about these sources of heat are based partly on what we think is the structure of the Earth and partly on how we think the Earth was formed. We think that the Earth was formed by the accretion of planetesimals (chunks of the original materials of the solar system) over four and a half billion years ago. As the Earth grew, its gravitational attraction also grew and the Earth captured even more planetesimals. In a short time the Earth was colliding with many objects of diverse sizes. Many of these collisions would have generated a good deal of heat,[2] as also did the compression of the accreting materials by the increasing gravitation. In the hot terrestrial interior, the heavier elements separated toward the center, eventually creating a metallic, radioactive core. Less heavy materials concentrated first in the mantle and then in the lithosphere (the crust and the uppermost portion of the mantle upon which the crust rests). The heat from the core stirs up the mantle and leads to the convection currents that push tectonic plates apart at the ridges. Through those long trenches, a new surface is forged from the mantle that spills forth.[3]

In trying to understand the mantle, the core, and so on, we do not observe them directly. We infer some of their properties from the measurements we make of the Earth, for example with seismic waves, using a technique called "seismic tomography," which is based on the notion that seismic waves travel at different speeds through different materials (such as molten metal or solid rock). And then we use those measurements to test certain theories, which in turn are used to discern the more fine-grained geophysical properties. Among those theories are our ideas of how a planetary body forms and how it distributes its energy. More specifically, they are theories of the evolution of a planetary body, adjusted to the Earth's mass and position within the solar system (not just its distance from the Sun but also its having a very large moon as well). Thus, to understand the Earth, we must understand what kind of planet it is.

Two Objections

A critic might raise two objections at this point. The first is that to understand the global environment of the Earth, we need at most to have some knowledge of the *present* structure of the Earth. We need to consider only the present mass and energy distribution of the Earth, not what happened billions of years ago. The second objection is that to understand the present structure of the Earth, we do not need to think of Earth as a planet. The structure of Earth does not depend on that of Mars or Neptune. Why then do we need to know how they are structured in order to know how the Earth is structured?

A simple consideration alone disposes of the first objection: the history of the Earth is important to determine its possible range of behavior in the future. Take as basic a matter as the age of the Earth. If the Earth is indeed four and a half billion years old, certain mechanisms are plausible candidates to account for the transformation of the environment. Plate tectonics needs tens of millions of years for some of the feats that we impute to it. Radical changes in the chemistry of the atmosphere (e.g., the rise in oxygen from a trace gas to a large component) might have taken bacteria tens, or perhaps hundreds, of millions of years. Imagine now for the sake of argument that all the evidence for the age of the Earth is wrong, and that the Earth is only six thousand years old. In that case, if the Earth formed roughly as we believe, it must have dissipated energy at such a high rate that the global environment must have been run by wildly different mechanisms. And since many of those mechanisms would be the same ones that operate today, or would have caused them, our understanding of today's Earth would have to be seriously mistaken. Thus, to understand the *present* global environment, and glimpse its future, we need to have some idea of how the Earth started and of how it evolved. That is not to say that without planetary science, including the evidence collected by the astronauts on the Moon, the only measure we would have of the age of the Earth would be the chain of "begots" in the Bible. Radioactive dating can be extremely accurate, of course, but if its use must be restricted to our own planet, then the results will be limited by the constant renewal of the Earth's surface.

As for the second objection, the structure of the Earth may seem independent from those of Mars and Neptune right now, but unless we reject the theory of planetesimals, Mars and Neptune form part of the context in which we understand the theory, and thus how the Earth came to be as it is. From them and other planets, we learn, for example, that a smaller, less dense Earth, or an Earth far closer to the Sun, might have defeated life's best efforts to gain a foothold and flourish.

The complex interactions between our planet's systems presently regulate in a fortunate manner our share of solar energy. But that energy does not remain constant. It appears that the luminosity of the Sun was much less during the first stages of the formation of the Earth, before its nuclear fires were ignited. And even afterward, the Sun's luminosity, according to some hypotheses, may have been 30% lower from what it is now.[4] The Sun also seems to undergo a variety of cycles in its output of energy. To complicate matters even more, the Earth's tilt with respect to the solar plane varies slightly (the spin axis of the Earth oscillates between 22 and 24.4 degrees every 41,000 years).[5]

The eccentricity of the Earth's orbit also changes slightly in cycles of 100,000 years (the orbit departs from its nearly circular shape). M. Milankovitch suggested many decades ago that this cycle was the cause of the Earth's ice ages, which also have a cycle of about 100,000 years. Since the two cycles could not initially be shown to coincide, and since no one proposed a generally accepted mechanism by which the expected change in luminosity would lead to an ice age, Milankovitch's hypothesis was met with skepticism.[6] Nevertheless, later studies of the history of the oceans provided strong evidence that the two cycles do coincide.[7] Of course, if variations in the energy output of the Sun, or in received luminosity, influence the Earth's climate, they will also influence that of other planets. We may then look in those worlds for evidence of such influence, and for a determination of the mechanisms by which that influence is exercised.[8]

It is also clear that the Earth is not a closed system with respect to matter. Given its position as one of the inner planets in the system, and its gravitation, the Earth attracts a good number of bodies, some of which collide with it. Most, if not all of these bodies, are debris left over from the formation of the solar system. They range from dust and small meteorites to comets and large asteroids. Such collisions may have a most profound effect upon climate and life.

Even in recent geologic times (within the last one hundred million years) large meteors have collided with the Earth, altered the weather catastrophically, and brought extinction to many species. One asteroid about 10 kilometers in diameter, now called the Alvarez asteroid, is held responsible for the disappearance of the dinosaurs about sixty-five million years ago.[9] Gravitational disturbances of the asteroid belt, the Kuiper Belt (a little beyond Pluto) or of the (possibly) billions of comets in the Oort cloud, in the outskirts of the solar system, will send several rather large bodies toward the Sun.[10] Some of them collide with the planets and moons of the solar system. In 1994, for example, large fragments of Comet Shoemaker-Levy 9 hit the atmosphere of Jupiter at velocities over 200,000 kilometers per hour, exploding with a brightness as much as fifty times that of the entire planet, and ejecting searing materials thousands of kilometers above the clouds. Had Shoemaker-Levy 9 hit the Earth instead, we would have gone the way of the dinosaurs.[11]

Apart from the realization that our natural history has to make conceptual room for such catastrophes,[12] there is a most obvious practical issue of survival involved. Perhaps with a reliable tracking system in place, *space technology* might allow us to change the orbits of those comets or asteroids most in danger of colliding with the Earth. But how worried should we be? According to present models, meteors large enough to create Meteor Crater in Arizona would hit an urban area every 100,000 years on average. That meteor was presumably 60 meters across; the crater is 1.2 kilometers across. A body with a diameter of 250 meters would cause a crater 5 kilometers across and destroy about 10,000 square kilometers (about the area of greater Los Angeles). Global catastrophes would take place every 300,000 years. These would be meteors with a diameter of approximately 1.7 kilometers.[13]

What is the evidence for these calculations? Soon after impact, craters are attacked by wind, water, life, lava, and a myriad of tectonic motions. In the blink of an eye, geologically speaking, all obvious traces of them disappear from the surface of our active planet. But we find a good record on the Moon. And in Venus, where most of the surface is 600 million years old, the spacecraft Magellan counted nearly one thousand impact craters at least twice the diameter of Meteor Crater. Venus is almost the same size as Earth, and in the Earth's vicinity, and since the impacts are geologically recent, it seems reasonable to fear a truly catastrophic impact on Earth every half a million years or so.[14] Whether those of us living today will experience such catastrophes, eventually our descendants will be thankful to us if we do create a warning system and the technology to deflect dangerous asteroids and comets.[15] There can hardly be a better reason than the preservation of life, and perhaps the survival of the species, to establish the importance of thinking of the Earth as a planet. This topic will be taken up more fully in Chapter 10.

The Moon also has a significant effect upon our global environment. In the short run the Moon affects the ocean tides; in the long run it slows down the Earth's rotation—one Earth's day may have been less than ten hours long. Just as the gravitational effects of the Earth on the Moon slowed down the Moon's rotation so that now the Moon always offers the same side to the Earth, the

Moon's gravitational attraction, though smaller, will eventually have a similar effect on the Earth. The night-day cycle is, of course, an extremely important component of our climate and presumably has played a major part in our natural history.

The Moon has also affected the climate in a second important way: its gravitational influence stabilizes the tilt in the Earth's axis of rotation, so that it varies only a few degrees. Mars, by contrast, may have suffered wild swings in the tilt of its axis, and this instability might have had devastating consequences for the Martian climate.[16] In other words, the Moon may have played a crucial role in ensuring that life on Earth endured and prospered while Mars became a barren world.

We have begun to see in this section that our specific theories about the Earth are inevitably tied to more general theories about the nature and behavior of the other bodies of the solar system. As we challenge our understanding of that system, we place ourselves in a position to learn new things, not only about other worlds but also about our own. The bounty of space science will thus not be scattered by alien winds over alien lands. It will be handed down to the children of the Earth.

The Exploration of the Solar System

The scientific exploration of the solar system provides rich support for the thesis that a better understanding of other worlds allows us to understand our own world better. In investigating other worlds, we find:

1. Valuable information that serves to refine our theories of the origin and evolution of the solar system, and hence of the Earth.
2. Unusual phenomena that stretch our views of basic terrestrial mechanisms.
3. Opportunities to test our ideas about the Earth—the solar system serves as a natural laboratory.

Valuable Information about the History of the Earth

The origin and evolution of the Earth are closely tied to those of the Moon. Until the advent of the space age, three main theories had been advanced to account for the origin of the Earth-Moon system. According to the Daughter theory, first proposed by George Darwin, son of Charles Darwin, the Moon was born of Earth material. Presumably some cataclysm caused a chunk from the Earth to go into orbit (Darwin speculated that the Earth tides formed by the Sun coupled with the free oscillations of a rapidly rotating Earth—every five hours—created a big bulge on the equator of the Earth, and that big bulge was thrown off).[17] According to the Sister theory, the Earth and the Moon formed side by side from planetesimals.[18] According to the third theory, the Wife theory, the Moon was simply captured by the Earth.[19]

A fourth theory, and the most popular view in recent years, is that a body the size of Mars collided with the proto-Earth.[20] In the ensuing explosion from this giant collision, materials from the two bodies were flung far and wide. The Moon accreted from materials that remained in orbit around the Earth. This explosion vaporized a greater proportion of silicates and volatiles than it did metals. The proto-Moon did not have enough mass to hold on to volatiles such as water, carbon compounds, and even some metals like lead, which means that silicates formed a large proportion of the Moon's materials. This result made the composition of the Moon very similar to that of the Earth's mantle in some important respects. The Giant Collision hypothesis thus explains not only the lower density of the Moon, but the abundance of silicates and the poverty of volatiles found by the Apollo astronauts.[21] Recent investigations by Junjun Zhang and his colleagues have determined that the isotopic composition of the Moon, pretty much like the Earth's mantle, shows little trace of the hypothesized Mars-size object. For some, that result implies that the Giant Collision hypothesis is false, but the authors themselves suggest several ways in which their findings are consistent with the hypothesis. For instance, the isotopic homogeneity of titanium, highly refractory element suggests that lunar material was derived from the proto-Earth mantle, an origin that could be explained by efficient impact ejection, by an exchange of material between the Earth's magma ocean and the protolunar disk, or by fission from a rapidly rotating postimpact Earth.[22]

Harold Urey, who won the 1934 Nobel Prize in Chemistry for his discovery of deuterium and later became one of the twentieth century's great figures in comparative planetology, helped persuade the Kennedy administration of the value of the scientific study of the Moon. Urey, who favored the Wife theory, thought that the Moon had already been formed when the Earth captured it, and that therefore it should hold valuable evidence of the early processes in the history of the solar system. But according to Urey's model, the Moon was already a cold body when the Earth captured it; the maria (the large flat areas that resemble seas) probably had formed when water splashed up from the Earth during capture; and, perhaps most important of all, the Moon's crust should have great quantities of nickel. The reason for this last prediction is that the Moon was not supposed to have an iron core. In the formation of a larger planetary body like the Earth, when the iron goes toward the center, it carries the nickel along. On the Moon, the distribution of nickel should thus be more uniform than it is on the Earth.

The astronauts' findings, however, made it clear that the Moon had been warm during its early history around the Earth, that the maria were made of basalt (probably the result of volcanism), and that nickel was not near the levels required by Urey's model.[23] A few years after men landed on the Moon, Urey gave up the Wife theory.

The clues astronauts found in the plains, craters, and crevices of the Moon about the forces that transformed it, and particularly the age and composition of the rocks they brought back with them, allowed us to challenge and replace our previous ideas of how planets form. According to a hypothesis first proposed in the early part of the twentieth century by T. C. Chamberlin and others, the solar system formed when a star passed too close to the proto-Sun. Since the Moon and the planets would have been born of the Sun, they would have been very hot and consequently their iron and other heavy metals would have collapsed into central cores. But if the solar system had been formed instead by the cold condensation of gas and dust into Moon and planets, only the more massive rocky planets like the Earth would have metallic cores. The evidence we found on the Moon thus played a part in the acceptance of the theory of planetesimals: grains of dust collecting first by intermolecular forces and then accreting by the action of gravity. It is from the perspective of this theory that theorists now explain the origin of the Moon as the result of a giant collision.[24] This theory also makes the best sense of the heavy bombardment of the solar planets by giant asteroids and other very large bodies. This bombardment should have been at its heaviest during the first half billion years of the formation of the solar system.[25] That is precisely the record that we have found on the craters of the Moon. The main contributions of the Apollo program were thus not political but scientific.[26]

Unlike the Earth, the Moon has neither atmosphere nor oceans and has not shown much geological activity for the past 2 billion years (see Figure 4.1). The record of the history of the solar system, let alone of the history of the Earth's immediate neighborhood, has therefore been preserved

Figure 4.1. The Earth-Moon system (see color plate). (Courtesy of NASA)

much better on the Moon. The oldest rocks found there are over 4.3 billion years old, and no rocks have been found younger than 3 billion years old.[27] On the Earth, on the other hand, the oldest rocks are 3.8 billion years old, and most of the surface (the bottom of the ocean) is only 0.2 billion years old or even younger. Thus, it is clear that the Moon can tell us more about the early Earth than the Earth itself can.

The Moon, however, cannot tell us the whole story, for its surface has not preserved intact the record of impact upon impact. First, meteors, large and small, have altered the surface of the Moon.[28] Second, the Moon must have had some internal heat, and perhaps some volcanism as a result. Although the Moon is less dense than the Earth, it presumably had its share of the same radioactive materials that exist in the Earth's core. The Moon's accretion, then, must have generated a good deal of heat also, although, again, much less than the Earth's.

This lunar heat would have dissipated at a faster rate than the Earth's heat, because of the Moon's smaller size. The reason lies in the ratio of volume to surface area. A larger planet has a *smaller* surface area relative to its volume. An increase in diameter increases the surface area by a power of two and the volume by a power of three (a doubling of the diameter leads to four times as much area and eight times as much volume; a tripling of the diameter leads to nine times as much area but twenty-seven times as much volume). If two planets have exactly the same amount of heat per unit volume, the one with the largest relative surface area will radiate away its heat sooner. The smaller body, in this case the Moon, will lose its heat at a faster rate. Moreover, as we have seen, the Moon had much less heat per unit volume than the Earth to begin with. Still the Moon's internal heat seems to have kept it somewhat active for over a billion years. That would have renewed the lunar surface to some extent.

In several respects, thus, there are limits to what the Moon can tell us.

To find a record that goes further back, we must look at smaller bodies in which the internal heating was negligible. The asteroids are good candidates, especially those in the main asteroid belt, between Mars and Jupiter. There is evidence that many asteroids underwent some thermal and chemical alteration about 4.6 billion years ago, but little since. Thus, they offer a record of some of the forces at work in the early solar system.

For a look at what might have been the original material in the solar system, comets are a good bet. Many of them have been under the influence of significant solar radiation for a relatively short time (sometimes only in the millions of years, whereas the planets and asteroids have been under it for billions of years now). In 2014, the European Space Agency's probe Rosetta placed a lander on Comet Churyulmov-Gerasimenko (Figure 4.2). There were communication problems with the lander, but future landings should offer rich information.

We can also find clues about the many factors that affect the evolution of a planet—internal structure, tectonics, or atmosphere—on most of the bodies of the solar system. Since they were formed under different circumstances— because of their position in the solar system and the distribution of materials in the sweep of their orbits—and since their interactions with the rest of the solar system are somewhat different from ours, they offer a wide range of instances of those evolutionary factors at work. It is not surprising that, under these different circumstances, unusual mechanisms have come into existence. In trying to understand such mechanisms, we modify our ideas about our own planet, as we will see in the next section.

Figure 4.2. Rosetta's image of Comet Churyumov-Gersimenko in 2014. (ESA/
Rosetta/NavCam—CC BY-SA IGO 3.0)

Stretching Our Views of Planetary Mechanisms

In the standard account of the evolution of a rocky planet, the denser the
planet, the more heat it will have available from radioactive elements; and
the larger the planet, the more retarded the loss of heat. By this account, no
planets much less massive than the Earth could still have active volcanoes or
relatively young surfaces. Mars, for example, shows evidence of recent volcanism (within the last two million years); but even if Mars is not a dead
planet, its surface is testimony to a prolonged coma. Nevertheless, Harold
Urey argued a long time ago for a greater variety of mechanisms that could
produce internal heat in a planetary body. Some theorists, following in
Urey's footsteps, went as far as speculating about volcanoes on Io, a Jovian
moon about the size of ours—an idea that seemed much too fanciful to most
researchers until, to their astonishment, they looked through Voyager's
camera and clearly saw the gigantic plume of a volcano rising over Io's horizon (see Figure 4.3).

Another surprise greeted them when Voyager discovered that the surface
of Enceladus, a small icy moon of Saturn about 1/100,000th the mass of the
Earth, looks quite new. And Europa, a beautifully smooth satellite of Jupiter,
apparently has large water oceans under a frozen surface. Evidence of early
geological activity can be found in many other moons, including Uranus's

Figure 4.3. Plume of erupting volcano in Io (see color plate). (Courtesy of NASA)

Oberon, Titania, and Ariel. One of the mechanisms that may explain these findings is that each of these moons is caught between two or more masses that exert significant gravitational attraction upon it. As a result of Io's specific position, for example, its mass expands and contracts in tides created by Jupiter on one side and Europa on the other. But we should not suppose that this form of tidal heating is the only additional mechanism able to produce an active geology. The bizarre geological formations in Uranus's moon Miranda can perhaps best be explained by supposing that Miranda has broken up one or more times and is now in the process of differentiation, with very large ice formations still side by side with big chunks of dark carbon compounds.

In Ganymede, the largest moon of Jupiter and the solar system, we can see what looks like signs of the beginnings of plate tectonics, now conveniently frozen for our inspection.[29] In other worlds we can see other stages of the generation and dissipation of internal heat. But none prove as instructive for the Earth as Venus.

Venus has an extreme environment and thereby presents a good opportunity to examine the relations between the climate of a planet and other systems such as its biosphere, its geophysics, and so on. The principle at work here is the same as in the rest of science; for example, a study under extreme temperatures may reveal the relationship between the fundamental physical forces, or extreme doses of a substance given to laboratory rats may reveal the possible action of that substance in the development of cancer in human beings.

Climatologists believe that living beings, the "biosphere," regulate in great measure the temperature of the planet by absorbing and releasing CO_2 and other greenhouse gasses, and that temperature is a crucial factor in the overall climate. And it is obvious that the climate has strong influence on those same living things. It also seems obvious that the climate has only a superficial effect on the geophysics of the planet, and thus it makes no sense to speak of a relationship between the biosphere and the geophysical structure of the planet. For example, neither the most powerful storm, nor a 50-megaton nuclear bomb[30] in our arsenal, is more than a mosquito bite compared to the planet's tectonic forces.

The exploration of Venus, however, might make us think about these matters differently, even though Venus, practically the Earth's twin in size and mass, has been extremely difficult to explore. The dense cloud cover keeps the surface hidden from our view. Part of the reason is that in Venus's atmosphere there is 300,000 times more CO_2 than in Earth's. The resulting greenhouse effect has helped produce a temperature of almost 900 degrees Fahrenheit (482 degrees Celsius), about the melting point of lead. In that oven, volatile substances are kept at a large height from the surface, where they form dense clouds that keep radiating heat downward (most sunlight is actually reflected by those clouds into space, which explains why Venus is so bright). The density is one hundred times that of the Earth's atmosphere, which turns a wind of 10 miles an hour into a hurricane. And whereas on Earth rain cleanses the atmosphere and changes the land, on Venus the rain is made of sulfuric acid and the heat evaporates it long before it can touch the ground.

In this inhospitable world, our landers perish in a matter of hours, unable to give us more than the vaguest of glimpses. This was the situation until the arrival of the spacecraft Magellan in the 1990s. Magellan's radar gave us maps of Venus better in many respects than those we then had of Earth.

Once upon a time Venus might have been very different. When the Sun was dimmer, oceans, rivers, and perhaps even life may have existed there. According to a plausible scenario, as the Sun became more luminous, life's regulation of carbon could not keep up with the increase in energy and a runaway greenhouse effect began to vaporize Venus's oceans. The increase in water vapor made the atmospheric temperature rise even more, which then vaporized more of the oceans. Eventually the oceans ended up high in the atmosphere, where ultraviolet radiation disassociated the H_2O to form atomic hydrogen (H) and the radical OH.[31] Most of the atomic hydrogen was lost to space while the OH entered into a variety of reactions with other substances in the atmosphere.

If anything like this scenario took place, one would expect a rather high ratio of deuterium to standard hydrogen. Normally only so many hydrogen atoms should be expected to be in the form of the isotope deuterium (deuterium has a neutron in the nucleus). But since deuterium is heavier, it is not as likely to be blown away from the planet; and thus, as time went by, it should have become a larger percentage of the hydrogen still found in the atmosphere of Venus. This is exactly the case.

In conversations with planetary scientists in the early 1980s, I floated the suggestion that the absence of water on Venus would change the viscosity of the rocks (viscosity is the resistance to flow) and, thus, plate tectonics was unlikely. The change in viscosity would make subduction (when a plate goes under another) and other plate motions very difficult. That is, Venus was unlikely to show much on the way of plate tectonics. Not to worry, I was told, high temperature can make the mantle behave like melting hot butter. By 1984, however, M. Carr and others had shown that lack of water would indeed make Venusian plate tectonics rather unlikely.[32] (I am sure they had been thinking along these lines longer than I and, moreover, had the expertise and imagination to come up with convincing explanations.) Most observers came to agree with Carr.[33] It seems, then, that a runaway greenhouse effect can deprive a planet of plate tectonics, and thus we should consider an astonishing corollary: without the climatic regulation by life, plate tectonics might have disappeared from the Earth as well.

This hypothesis, which seemed so fanciful forty years ago, is apparently considered quite reasonable nowadays. Indeed, the reasoning that takes us to the biological modulation of plate tectonics goes further. As the planetary scientist D. H. Grinspoon puts it:

> If you agree to that, you must agree that Earth's interior thermal evolution has been affected by its changing atmosphere and biosphere, because plate tectonics is the main way that Earth cools its interior. Even such remote quarters as the molten iron outer core, which produces Earth's singular magnetic field, may not have been immune to the modifying effects of Earth's quirky air, its unique, biologically touched, gaseous envelope.[34]

Nevertheless, the hypothesis that Venus once had oceans is by no means universally accepted. For it depends crucially on the plausible reading of the ratio of deuterium to common hydrogen as evidence that Venus has lost a substantial amount of water. But this is not the only possible reading.

Some scientists have argued that the same ratio of deuterium to hydrogen could be caused by a continuous resupply of water to the atmosphere of Venus. They find this hypothesis more plausible because, at the present rate of escape, water would disappear from Venus altogether in a few hundred million years. This would mean that right now we are witnessing the very end of a long process. Most of us would not be bothered by this prospect, but some people feel that we should treat with suspicion all lucky coincidences in science.

The intuition behind this reasoning is a peculiar view of probability that sounds quite reasonable at first. Suppose that an urn contains two white balls and 98 black balls, while another contains two white balls and 9,998 black balls. Without knowing which urn is which, you reach into one of them and come up with a white ball. You should conclude that in all probability you have reached into the smaller urn, for the chance of getting a white ball is 2 in 100, which is much higher than the 2 in 10,000 that you would find in the larger urn. It is not reasonable to believe that you have reached into the larger urn, since getting a white ball out of it is so much more unlikely.

These intuitions about probability come with their own problems, however. The first problem is that it turns scientific reasoning on its head. In the case of Venus, it leads us to expect that the amount of water we find is normal (that it has been like that for a long time). It leads us to assume that things are pretty much as we find them (a steady state) because otherwise the situation would be very unusual and thus unlikely.

Notice how differently we reason in science. When we first observe any phenomenon, we take pains to determine whether our sample is representative. The history of science is littered with ideas that seemed promising but went nowhere because they were based on the assumption that we were looking at the normal state of things (this is the fallacy of induction). That is, normally we *have* to demonstrate that our sample of observations does represent the usual state of affairs. We are not allowed to take that for granted. Otherwise, we may conclude, say, that we have discovered a new branch of the homo family based on one fossil with a peculiar skeleton, only to find out much later that it was the skeleton of an individual with a bone disease (a true case, by the way). We worry about whether the *Viking* and *Venera* landed in representative locations in Mars and Venus, and whether the Galileo probe went into a section of Jupiter that is like the rest of the atmosphere (it didn't).

Nature is rich and what we come to observe may actually be as unusual as flowers, birds, and bees are in the solar system: they exist in only one world—Earth—out of the many that orbit the Sun.[35]

The kind of reasoning that leads to a steady-state view of water in Venus also leads to some very strange views when applied elsewhere. John Leslie, for example, has concluded in *The End of the World* that the human species is likely to become extinct very soon.[36] He reasons that if the human species were to live for a long time, let's say millions of years, then the amount of people alive today would be an insignificant percentage of the total amount of human beings that will ever be alive. Thus, he thinks, belonging to such an unusual group of humans (those of today) would be extremely unlikely. On the other hand, if the world were to end within fifty years or so, we would be part of the largest group of humans who will ever be alive (the seven and a half billion or so alive today are far more than all the rest of the humans who have ever lived put together). And it is more likely, then, that if you were to pick a human at random, he would belong to the overwhelming majority than to a very small minority. Therefore, the human species is far more likely than not to become extinct very soon.

Many sensible people would consider Leslie's argument a *reductio ad absurdum* of the kind of probabilistic reasoning under examination here. Nevertheless, he, like others, sticks to his probabilistic intuitions, despite counterarguments like the following. Suppose the devil places ten people in a room and tells them that he will kill them all if he gets double sixes in a roll of the dice. If they survive, he will then place 100 people in the room, and then 1,000, and so on, always multiplying by ten. And every time the devil will roll the dice in hopes of getting a double six. Now, most of us will think that the chances of any one group getting killed will be 1 in 36, but according to Leslie, the ill-fated chances of the group of 10,000 have to be far greater than those of the group of ten. Leslie concludes not that there is something very wrong with his reasoning, but that he has encountered a paradox of probability. And he remains worried about the end of humanity, just in case the paradox should resolve itself in the direction of his reasoning.

Probability without Paradox

Leslie's intuitions about probability may remind us of the double lottery winner. Suppose that you buy lottery tickets for Tuesday and Friday. There are ten million possible numbers each time, so your chances of winning on Tuesday or Friday are each one in ten million. And now suppose that you win on Tuesday. What are your chances that you will win on Friday also? Some may feel that the chances of winning on Friday after having won on Tuesday must be far less than your chances when you won on Tuesday: some calculate one in ten million times ten million ($1/10^{14}$). But this is a mistake. It is one thing to ask, what are your chances that you will win on both Tuesday and Friday? The answer is $1/10^{14}$. It is another thing to ask, what are the chances that you will win on Friday *after* winning on Tuesday? Well, you have one ticket and there are ten million possible numbers. So your chances should be one in ten million (10^7). But people like Leslie find this hard to accept. They think that there is something valid about the intuition that tells us that after winning the lottery once the chances of winning it again have to be far smaller than they were the first time. It would be too much of a coincidence otherwise.

To dispel this intuition, let us look at one example in which the probability grows with every independent and successful step toward a result that would be highly improbable if we were to calculate a conjunction of all the steps. Suppose you flip a fair coin a hundred times. What are the chances that you will get a hundred heads? $1/2^{100}$. That is an extremely small number. Suppose now that on the first flip of the coin you get heads. What are the chances now that you will get 100 heads? There are 99 flips left. That means that you will have to get heads on all 99. Your chances then will be $1/2^{99}$. That is a very small number, too, but it is twice as large as the number you had before (because the denominator, 2^{99}, is half of 2^{100}: 2×2^{99} is 2^{100}). Confirm it: $2^{99} + 2^{99} = 2^{100}$. Consider two easier examples: $2^2 + 2^2 = 2^3$ ($4 + 4 = 8$) and $2^3 + 2^3 = 2^4$ ($8 + 8 = 16$.) If the base is 3, then you have to add three numbers: $3^2 + 3^2 + 3^2 = 3^3$ ($9 + 9 + 9 = 27$). If it is 4, then four numbers, and so on.

After getting 2 heads, your chances of getting 100 heads will be $1/2^{98}$, which is a very small number, but twice as large as that of your second flip and four times as large as your first. The more heads you get, the more your chances improve. Suppose you have flipped 98 heads, so you have only two flips left. Your chances will be $1/2^2 = 1/4$. You have gotten 99 heads? Your chances of getting 100 heads depend on your getting heads on your last try. That is 1/2. Of course, the probability of getting 99 heads and one tail at that stage will also be 1/2.

As we have seen, then, your chances of getting 100 heads when you have already gotten 99 is far greater than when you had gotten your first head, and greater still than when you had not yet begun to flip the coin, while your chances of getting heads in your last flip are exactly the same as the same as in your first flip: 1/2.

There was never a paradox. Some probability estimates are more relevant than others. Suppose that a ninety-two-year-old cancer patient goes to the hospital to have an appendectomy. He is informed by his doctors that 98% of all patients who undergo the operation do well. Unfortunately, cancer patients in his age range survive less than 5% of the time. Which probability estimate should guide his decision to undergo the operation? Similarly, against Leslie, the relevant probability is that the devil will kill all the people in the room if he rolls double sixes.

It may turn out that the amount of water of Venus is in a steady state, but to have confidence in that idea we need independent scientific reasons in its favor, not peculiar intuitions about probability. A candidate to keep the Venusian atmosphere resupplied with water is volcanism. A good possible outside source of water is the combination of comets and their fragments, for they are basically a mixture of ice water and other compounds. As Grinspoon points out, however, comets collide with planets often but not continuously and, thus, we would not know whether that hypothesis is consistent with our present readings of Venusian water (comets would supply water in spurts). Furthermore, both this hypothesis and that of volcanism assume that the exceptionally large ratio of deuterium to hydrogen would exist in Venus at all times. But this assumption is implausible, for the present ratio is 120 times greater on Venus than on Earth! It is difficult to imagine what could account for such a phenomenal difference in natural ratios between the two planets. The original ocean hypothesis must continue to be considered most reasonable until further notice.

Mars, by contrast, did not have as much internal energy as the Earth because it was smaller and less dense. And since it is further from the Sun, it receives less sunlight. To compensate, if Mars were to have as comfortable a climate for life, it would need a much greater greenhouse effect than the Earth. This would require truly large amounts of CO_2 in the atmosphere. At the present time the CO_2 in Mars is fifty times per unit volume that of the Earth's atmosphere. That sounds like a lot, but it produces only a puny greenhouse effect. The reason is that since the atmosphere is already so cold and thin, the Martian water is frozen at the poles, visible at the north pole, and under the frozen CO_2 sheet at the south pole. It is also spread as permafrost under large areas of the surface. In addition, striking discoveries have been made by the Mars Advanced Radar for Subsurface and Ionosphere Sounding (MARSIS), particularly a rather large underground lake (about 19 kilometers wide) near the south pole and three additional smaller, lakes, all of which seem to contain very salty liquid water.[37] This suggests that much more water may be found even deeper underground. Just the ice detected on Mars so far amounts to about 5 million cubic kilometers of water, which, if melted, would suffice to cover the whole planet with an ocean 35 meters deep.[38] It is frozen, though, and this means that the initial boost that CO_2 gives to the greenhouse effect is not multiplied by that of water vapor, unlike Venus or Earth.

It seems, however, that earlier in its history, when Mars's internal energy was much higher, Martian volcanoes might have filled the atmosphere with as much as one hundred times as much CO_2 as today.[39] This factor would have raised the density and temperature enough to permit liquid water and large amounts of water vapor (see Figure 4.4) and hence much more of a greenhouse effect.

This view of Mars is strengthened by indications that Mars at one time did have some form of plate tectonics as well and may still have some remnants

Figure 4.4. History of water on Mars. Numbers represent how many billions of years ago (see color plate). (Courtesy of NASA/Michael Carroll)

of it in the Valles Marineris.[40] In the face of all these considerations, it may seem reasonable to imagine that life could have existed on Mars. If so, why did not Martian life control the climate the way Earth's presumably did? Mars apparently did not have enough energy to run the cycles that have made a sustainable biosphere on Earth possible, just as it did not have enough heat to support the motion of tectonic plates for billions of years. In addition, Mars lacks a magnetosphere, which means that the atmosphere is not protected from the solar wind and is lost to space as a result. Life, if it ever existed on Mars, was thus powerless to stop the ultimate collapse of its global environment. Earth was fortunate not to be besieged by the extreme conditions of its two sister planets, as we will discuss shortly.

Understanding the Global Environment of the Earth

CO_2, Climate, and Other Global Problems

Carbon dioxide (CO_2), water (H_2O), and sunlight are the most crucial ingredients used by plants and many bacteria to make the organic compounds they need to survive and prosper (through photosynthesis). CO_2 is thus essential to life on Earth. But in excessive amounts its greenhouse effect may lead to the infamous "global warming."

The theory of the greenhouse effect seems solid. Sunlight is absorbed by the Earth's surface and re-radiated in the infrared. CO_2 traps infrared radiation, as do other "greenhouse" gases, such as methane (CH_4), nitrous oxide (N_2O), and ordinary water vapor;[41] and this energy makes the atmosphere warmer. Indeed, the Earth experiences a large greenhouse effect. The average air temperature on the surface of the Earth was 14.9 degrees Celsius (58.6 degrees Fahrenheit) as of 2017. Back in 1950, it was 14 degrees Celsius, about 32 degrees Celsius higher than the radiant temperature of the Earth (what the Earth radiates). This means 32 degrees Celsius more than can be accounted by the combination of sunlight and internal heat. So we can see not only that the greenhouse effect exists on a large scale, but also that it is *generally* a good thing; otherwise the mean temperature of the Earth would be well below freezing.

Even though the percentage of atmospheric CO_2 is just 0.04, its increase over the last 120 years (from about 0.03) presumably has led to increases in temperature that have in turn led to increases in water vapor, and thus to still higher temperatures. At first sight, however, it seems that, through this multiplier effect, *any* increase in CO_2 would inevitably lead to a runaway greenhouse effect. But then the Earth's temperature should already be measured in the hundreds of degrees Celsius, as in Venus, for surely there have been fluctuations in the level of CO_2. Indeed, such level has apparently been much higher in the past: ten to thirty times higher in some eras, and also a bit lower than it is now (two-thirds of today's level during the last ice age).[42] Ninety million years ago, a cold-blooded crocodile, Champosaur, lived only 600 miles from the North Pole. The climate in the Arctic then was like Florida's today.[43] But clearly the Earth is not lifeless.

Hence our planet must have mechanisms that regulate the level of carbon and its effect on temperature. Three mechanisms seem particularly influential: plate tectonics, stone weathering, and terrestrial life.

Let us begin with life. As CO_2 increases, plants and bacteria that thrive on it will displace others that do not. But as these life forms become very successful and proliferate, there will be more of them to remove CO_2 from the atmosphere, and thus the temperature will begin to go down.

Lynn Margulis and James Lovelock predicted that, in addition to plants, bacteria are also an important biological sink of CO_2.[44] The Sea Wide Field Sensor satellite, launched in 1997, showed that plants removed about 52 billion metric tons of inorganic carbon, and plankton as much as 45 billion to 50 billion metric tons.[45] We should see this pattern repeated time and again: firm knowledge about problems of the global climate seems more likely when we can investigate them with space science and technology—space technology is of course particularly relevant to our interests.

Plants, however, contribute to the removal of CO_2 not merely by photosynthesis but also as part of the so-called silicate-carbonate geochemical cycle, which works by taking the calcium living beings produce and combining it with carbonic acid to make limestone. As astrobiologists Peter D. Ward and Donald Brownlee explain:

> Here we have a wonderful partnership. Animals such as coral are harnessing calcium. The roots of plants exude an acid that helps to break down rocks, accelerating weathering by the wind and rain generated by the atmosphere and oceans, creating the [carbonic] acid necessary to convert the calcium to limestone. All combined are working together to take excess carbon dioxide out of the atmosphere and bury it in "reservoirs" of rock within the Earth, and thus balance temperature.[46]

To understand the regulation of CO_2, we also need to determine the ways in which CO_2 is put back into the atmosphere. The oxygen produced by the organisms that remove CO_2 is itself taken up by other organisms, which end up producing more CO_2 as waste. Plankton stores carbon in the ocean, but some of that carbon is returned to the surface via upwelling and ocean currents in a few hundred years.[47] Most of that carbon, though, becomes part of the crust, as carbonates, and because of the spreading of the ocean floors through plate tectonics, is eventually pressed onto continental shells as plates collide (in subduction zones) and finally finds its way into volcanic eruptions as CO_2 again. The lag in this geologic cycle may extend into the millions of years.

When the planet begins to cool, more water freezes in the polar caps, the level of the oceans drops, and more land is exposed to the wind and the rain. Phosphates and calcium in great abundance come to the oceans. The phosphates feed the plankton, which will then make more carbonates from atmospheric CO_2. As a result, the temperature will decline even more.

As the ocean surface is reduced, the ability of the planet to support plankton is also reduced. A saturation point is eventually reached, and the temperature becomes stable. After a while the combination of plate tectonics and volcanism will begin to increase the level of CO_2 and the temperature will begin to rise again (although big volcanic eruptions put up large amounts of dust that initially may cool the planet more instead).

In presenting his famous Gaia hypothesis, James Lovelock has compared the totality of Earth's life, the biota, to a super-organism with a fierce instinct for self-preservation. As his metaphor is pressed to explain the natural history of our planet, it does become clear that life has always managed to adapt to profound changes in the environment. It also becomes clear that the biota is a very effective mechanism in the regulation of that environment.

For example, Lovelock points out that the present industrial pollution of the planet cannot compare with the massive poisoning of the atmosphere by oxygen about two billion years ago. When the free hydrogen in the atmosphere had been used up or escaped into space, bacteria began to withdraw it from H_2O in photosynthesis. The waste product, oxygen, was extremely toxic to most of the bacteria that dominated the Earth in those days.[48] Nonetheless, life overcame this threat by developing organisms that used oxygen. But we should find no comfort in learning that those evolutionary solutions led to the replacement of one kind of life by another. If we foul up our own planet, life may survive—but we might not. Industrial waste fertilizes the water in our lakes and rivers and leads to the rapid growth of algae. This kills the fish, the frogs, and the water lilies. It is not much consolation to know that upon the ruins of the present order life will adapt and produce a new kingdom of scum.

Life is so effective a regulation mechanism that it maintains the atmosphere far from chemical equilibrium. For example, oxygen forms 21% of the volume of the atmosphere. From a purely chemical point of view this high percentage is very surprising, for oxygen is a very reactive element (it combines easily to form compounds); thus, it should be swept up in a rather short time. Life, however, replenishes the free molecular oxygen that is lost to chemical reactions. And given the large amount of oxygen, the percentage of other gases would be impossible except for the action of life. Methane, for instance, is 10^{29} times more abundant than it ought to be, nitrogen 10^9 and nitrous oxide 10^{13}, given chemistry alone.[49] If a spaceship were exploring our solar system, it might be able to determine the existence of life on Earth, long before arriving, from the extremely high concentrations of trace gases such as methane and nitrous oxide.[50] The converse is also true. By spectral analysis, Lovelock determined in the early sixties that life on Mars would be very unlikely.[51]

The concern of climatologists today is that, because of our increasing use of fossil fuels, we are outstripping our environment's ability to cycle CO_2 properly. The resulting global warming later this century will bring us coastal cities flooded out of existence, large areas of farmland turned into deserts, and, in short, a world under stress. They thus urge us to think about the Earth in the long run, meaning the next few centuries. Before discussing this scenario, I want to emphasize that I do not wish to discuss the controversy now called "climate change" here. That deserves a book of its own. I merely wish instead to do two things. One is to bring up some findings from space science that perhaps should be taken into consideration if we are going to take a thorough look at the problems we face. And the other is to point out that when trying to collect the most relevant facts or to put into effect potential solutions, space technology may prove very helpful.

Some space scientists have placed their own concerns about Earth's climate in the context of the truly long run, not just centuries, but thousands of years, and then tens and hundreds of millions of years.

For the past 400,000 years, according to the temperature record found in deep ice in Antarctica, our planet has been going through glacial cycles of about 90,000 years, punctuated by short warm interglacial periods (about 11,000 years each). Indeed, our planet has been very cold for the last 2.5 million years and is likely to remain so for the next 2–10 million. As Peter D. Ward and Donald Brownlee point out in their book *The Life and Death of Planet Earth*, the present interglacial period has been unusually long already[52] and will give way to another glaciation in a few thousand years at most,[53] although, as the record shows, the switch from warm to ice can take place almost suddenly. Consider then that the descent of sheets of ice two miles high upon the Northern Hemisphere would devastate our present civilization far more than the current warming of the atmosphere is likely to. This would be compounded, of course, by a similar disaster in the Southern Hemisphere. Ironically, according to some, the much-maligned human-made global warming may stave off the ice for another 50,000 years![54]

In the truly long run, but long before the Sun becomes a red giant, the Earth's "thermostat" is likely to malfunction. It was Lovelock himself who realized that Gaia would eventually fail as the planet's self-regulatory mechanism. Life has been steadily removing CO_2 from the atmosphere for the last 400 million years, when plants conquered the land, and in about 100 million years the level of CO_2 will go below 150 parts per million (ppm) of air.[55] This is the level of CO_2 most plants require to survive. Newer forms of plants— grass, palm trees—use slightly different mechanisms for photosynthesis and can go well below the 150 ppm. If this view is right, the flora of the future, then, will have a very different view: gone will be the apple orchards and the rose gardens, replaced by new and exotic varieties of plants. But, eventually, the level of atmospheric CO_2 will go below 10 ppm and photosynthesis will come to an end altogether. More recent studies offer a revised estimate of between 500 million and a billion years.[56]

The loss of plants will be a catastrophe for animals, obviously, but also for marine life, since it depends so much on the run-off of the soil nutrients that results from the presence of plants. Those few animals that can manage to survive will be obliterated in a few million years by the rising temperatures, for eventually significant levels of atmospheric CO_2 will rise in the atmosphere by geological processes but will no longer be kept in check by (extinct) photosynthetic organisms. Several scenarios have been proposed to predict what will happen after that point, but it seems plausible that a highly increased level of solar energy coupled with high levels of atmospheric CO_2 will quickly lead to the sort of runaway greenhouse effect that vaporized Venus's oceans.

It is possible, nonetheless, that Lovelock and those who refined his pre-diction of doom have not given Gaia enough credit. After all, just as some cyanobacteria were able to survive in small, protected pockets from the poi-soning of the planet by oxygen, a few plants may just barely survive near vents that outgas CO_2, lie low until the CO_2 rises again, and then explode once more through ocean and land. Other photosynthetic life on land and in the ocean will thrive also, as their ancestors now do, and together with the plants will begin to regulate the climate again and, literally, give the Earth a new lease on life.[57] That lease will run out in four or five billion years, when the Sun will become a red giant, overwhelming, and finally scorching into oblivion all: life, oceans, and air.

Creative Solutions to the Problem of CO_2?

Most climatologists presumably believe that factories, cars, and other human activities will raise the average global temperature anywhere from 1 to 5 degrees Celsius later this century.[58] Nevertheless, as mentioned earlier, for the sake of thoroughness we may wish to consider some scientific views that place at least some of the blame for the rise in temperature on factors other than human action. The role of the Sun, for example, may be quite surprising. Beginning in the 1990s, and coinciding with a maximum in Total Solar Irradiance, the ice caps in Mars shrank, in synchrony with the Arctic ice caps on Earth. As the Sun cooled in recent years, the ice caps in both planets began to return. This extraordinary synchronicity in the two planets across the vastness of space can only be ascribed to the Sun.[59]

In a different example, Jan Veizer and Nir Shaviv have argued that cosmic rays are a likely driver of temperature changes and claim that "The global climate possesses a stabilizing negative feedback. A likely candidate . . . is cloud cover."[60] And Henrik Svensmark at one point suggested that the solar wind partially shields our atmosphere from cosmic rays; thus, when the solar wind is low, more cosmic rays interact with water molecules, leading to cloud formations that reflect sunlight, which leads to a cooling effect. The opposite process has a warming effect.[61] Svensmark's insight was supported by an experiment that mimicked the interaction of cosmic rays with Earth's atmosphere—an experiment conducted at the Danish National Space Center.[62] Years later he continues his investigations, now in the field, so to speak. In a study published in 2016, he examined the data sets from three satellites and one ground-based to conclude that decreases in the galactic cosmic ray radiation indeed match their expected impact on the ionization of the lower atmosphere. These results demonstrate that there is a real influence of such decreases on clouds, probably through ions,[63] leading to an increase in atmospheric temperatures.

Moreover, nearby supernova explosions lead to large short-term increases in global cosmic rays, and the geological record for the past 510 million years does indicate a high correlation between local supernova explosion rates and a cold climate on Earth.[64] The emphasis on how the increase of ionization as a mechanism for the growth of aerosols into cloud condensation nuclei is an interesting way to meet the requirement of a natural explanation for the historical and archeological evidence of global warming and cooling periods before the Industrial Revolution, as Svensmark emphasizes. At any rate, this passage gives just a glimpse of one very interesting scientific line of thought.

Of course, we can understand why those scientists who foresee a catastrophe vehemently disapprove of delays. By the time we have conclusive proof, they say, it may be too late. From a professional philosophical perspective, it appears at first sight that they are right: under conditions of uncertainty, it seems reasonable to move so as to avoid disaster.

On the other hand, there are some who suggest that we should investigate alternative solutions even if the problem is truly serious. Long ago already, a National Academy of Sciences panel (1992) suggested that we should find means to compensate for any undue increase in greenhouse gases. The physicist Gregory Benford has cataloged several such means and has devised more of his own. One simple way to remove a large portion of the increase in CO_2 is to plant more trees. A campaign of reforestation in unused lands can be done for a few billion dollars, enhancing the quality of life, and giving us some breathing room in which to think of longer-term solutions.[65] As it turns out, however, many reforestation projects have failed. Different areas have their idiosyncrasies, and it is important to determine not only the right kinds of trees to plant in those locations but the right combination of trees and crops or grass to go with those trees. For example, in Africa, where 60% of agricultural land is degraded, this kind of regional approach seems very promising. And elsewhere this may also be the right way to turn near deserts into green lands full of trees, while allowing the local people to live off the land.[66] Clever LANDSATs may help alert us to projects that are not going well early and to search effectively for alternative approaches.

We need to know more about how the global and regional environments interact. Think about how irrigation and other uses of land affect the local climate. Irrigation in the plains of Colorado, where corn and other crops have replaced dry prairie, has reduced the mean temperature in July by 2 degrees Celsius. Transpiration by plants seems to cool the air and produce clouds over the plains. Further irrigation, then, may enhance the effect, for the plains' winds will cool nearby mountains, and that will increase rainfall there.[67] On the other hand, as we have seen earlier, limits may be encountered unless the proper kind of forestation (i.e., fruit trees) in the right amounts and places is made part of the project. And we should consider the possibility that the global problem of climatic change will largely resolve itself into a collection of regional problems. We would then still have to determine how those regional climates interact with each other and with the global climate. Weather and land satellites may prove very useful in this task.

Plankton can provide another solution, since they can withdraw a considerable amount of fossil-fuel emissions as well. Plankton are scarce in the polar oceans, in spite of the presence of large reserves of the nitrates and phosphates that plankton normally use. The reason seems to be the absence of iron in those oceans. But surely, John Martin suggests, we could ferry the needed iron dust for a relatively low price. According to Benford, for about $10 billion a year we could farm enough plankton to absorb as much as a third of a year's fossil-fuel emissions.[68]

Environmentalists such as Bill McKibben, on a mission to decrease the use of fossil fuels,[69] might wish to give Martin's idea some consideration. In a week-long experiment off the Galapagos Islands in 1996, 990 pounds of iron made the waters bloom with plankton, Benford reports, which then covered 200 square miles, suddenly green.[70] This was just one of many successful experiments that were carried out from 1993 to 2002. Some oceanographers worry that long-scale fertilization over a period of many decades may adversely affect the ocean.[71] But the iron dumping, at least, is a reversible experiment; for if the iron dust is not continuously supplied, the polar ocean will return to its previous state in ten days or so.

Other possible solutions involve changing the albedo of the planet: if it reflects more sunlight, it will absorb less energy, and therefore the greenhouse effect will diminish, other things being equal. We might achieve this goal by increasing sulfur emissions in the South Pacific (which would increase cloud cover over the middle of the Pacific Ocean, the darkest and most solar-energy-absorbing area of the world). Or we might, alternatively, make jetliners burn a richer fuel mixture, which would leave a trace of fog high in the atmosphere. According to Benford, this process would offset a year's worth of U.S. emissions for a mere $10 million!

These are suggestions. Readers may worry about the possibility of hidden long-term environmental costs and dangers of the suggestions that involve geo-engineering. Eventually, the story will likely be resolved by space technology used in combination with fields like oceanography and paleobiology. The idea behind the "Mission to Earth," for example, was precisely to look at Earth from space the way we have looked at other planets. This most ambitious mission evolved into the somewhat more modest Earth Observing System, a multi-billion-dollar interdisciplinary effort to understand global climate by observing the Earth from space. The three main instrument platforms are called Terra, Aqua, and Aura (to study, respectively, land, water, and atmosphere), and other satellites have been added or are planned, such as the high-resolution Sentinel 6, launched in November 2020, that should be able to read sea levels very precisely (in concert with its future companion, Sentinel 6B, to be launched in 2025). Nevertheless, it is clear to many scientists that even this effort must be surpassed. A recent meeting of the National Research Council attempted "to do for climate change what has been done for astronomy, planetary science, and solar physics: create consensus on a realistic, long-term blueprint for the field, including the most important questions to be answered and the tools needed to explore them."[72]

The most fascinating proposal to having electricity without fossil fuels remains the proposal to use solar power satellites for clean, abundant energy, discussed in Chapter 2, particularly in O'Neill's version. The materials will come from lunar regolith (8% aluminum and an abundance of other metals plus silicon). This lunar dirt will be placed in the buckets of O'Neill's mass driver, accelerated to a high velocity that will put it in lunar orbit, and particularly into the cone of a Catcher. Once there, the aluminum and other metals will be separated and processed through O'Neill's extremely ingenious building machines to make the reflectors, which will then go into geosynchronous orbit. This approach would be far cheaper than rockets. O'Neill made a plausible case for the ability to build reflectors the size of Manhattan Island. But technological advances have reduced the size of a reflector to that of a football field. Thus, the effort to produce one solar power satellite then can now yield probably most of the solar power satellites the world needs. Both the Japan Aerospace Exploration Agency (JAXA)[73] and NASA are considering building such satellites in the 2030s.[74] It will be interesting to see to what extent they follow O'Neill's model.

Another fascinating proposal is the solar shield. The concept is to place in the Sun-Earth Lagrange equilibrium point L1, which is always in the line between the Sun and the Earth, a shield that would reduce the amount of solar radiation, thus reducing the temperature of our planet as well.[75] The shield would actually be a collection of shields that would be arranged in L1 so as to fine-tune the amount of solar radiation we wish to receive. At the present time, the proposal is not in favor because it would take an enormous effort to build the shields, place them in orbit, and then deploy them. It would be far too expensive and might have a very drastic impact on the environment.[76] But if you were to build them from lunar materials and in lunar orbit, following O'Neill's suggestions for his solar power satellites, all these problems would be considerably reduced.

Of far greater importance in the long run may seem to be the ability that space technology may give us soon to protect our planet from collisions with asteroids, collisions that may range from the destruction of large cities to the extinction of human beings and most other terrestrial life. This topic will be taken up later, especially in Chapter 10.

We, of course, face other more immediate global problems—for example, pollution. Perhaps the most worrisome pollution is that of plastic, which in the form of small particles (microplastics) has invaded our oceans and is harming fish and other marine life. On this issue we have some help from space technology: a system of eight weather microsatellites (CYGNSS) has been used to identify concentrations of microplastics in the world's waters. These visualizations make the cleaning efforts by ships and other resources far more effective.[77]

This section has provided many examples of actual and potential benefits that space technology can offer us in our efforts to preserve and improve the lot of humanity. It is not intended, however, as support for some rather incautious promises to the effect that an extraordinary abundance of resources from the solar system may soon be bestowed upon the inhabitants of this planet. Except for plentiful clean energy from solar power satellites, such promises have not been the intent of this section. It is one thing to say that once humanity becomes a space faring civilization, it will find overwhelming treasures in the solar system. It is another to end up suggesting that the rather limited resources to be found in the Moon and near-Earth objects are also boundless treasures soon to be enjoyed on the surface of our planet. For lunar resources, consult Ian Crawford,[78] and for a painstaking argument against these perhaps misleading promises, see James S. J. Schwartz.[79]

The Solar System as a Laboratory

We often extrapolate with profit the bits and pieces we learn about the Earth to our understanding of other planets. Ideas about aspects of the Earth do serve as the basis for hypotheses about what we may find elsewhere. But this ability only strengthens our dependence on planetary science. For since those ideas about the Earth are also ideas about how a particular planet behaves, their worth is often appreciated only by seeing how they apply to *other* planets. Thus, the solar system provides us both with an appropriate context in which to interpret our observations of the Earth and with valuable opportunities to test our ideas about the Earth.

Furthermore, in order to understand the evolution of the Earth, we need to know what factors originate and shape planetary environments. But it is difficult to determine the range of those factors when we only see them in operation on the Earth. Although, as we saw earlier, we can carry out some experiments on the Earth's weather (and perhaps on its geophysics), the range of global controlled experiments is likely to be very restricted for a long time to come. That is a problem, for in science when we have an idea, we want to see how it works; we want to look at it from different angles. Fortunately for us, the solar system can take the place of the laboratory. If we want to know how the mass of a planet influences its tectonics, we look at several planets with a variety of masses and examine their tectonics. We cannot vary the global conditions of our planet at will. But we can look at other worlds in which those variations occur naturally and see how other factors are correlated with them.

On Earth, the oceans provide a buffer for heat and supply most of the water vapor in the atmosphere. What other influences are there on planetary weather? The study of planets without oceans begins to answer that question. How much of a factor is the planet's rotation? Jupiter, with its gigantic, three-layered atmosphere, rotates every ten hours. In Venus, by contrast, the day is the equivalent of 243 Earth days (the atmosphere, however, rotates sixty times faster than the planet). Computer models of weather systems that control for this and other factors have been tested by the actual performance of the atmospheres of the other planets. This gives us to some degree the analog of manipulating our own atmosphere to test our ideas about the Earth's weather.

For example, on Venus's dense atmosphere the temperature variations are minimal, 10 to 20 degrees Celsius, between poles and equator. This characteristic surely contributes to the absence of some of the weather patterns familiar on Earth. On the other hand, on Mars, the thin atmosphere does not offer resistance to atmospheric waves, with the result that such waves do not form eddies. The result is that on Mars the weather patterns are far more regular, repeatable, and periodic than on the Earth. Since the structures that organize the Earth's weather are not periodic, our weather is very difficult to predict. Nevertheless, we can hope that the comparative study of planetary atmospheres will continue to identify the factors that contribute to the behavior of our own weather.

The transformation and fine-tuning of our views about the Earth are by no means the only benefits likely to accrue from the exploration of the solar system. We will also increase the precision with which we can observe and predict a variety of environmental states. This increase will come, partly because we will be better able to specify relevant factors and parameters of the global environment, as a result of our comparative study of the gravitation, magnetism, atmospheres, morphology, topography, geology, and chemistry of the Earth and other planets; and partly because of the advances in the technology necessary to carry out that comparative study. At the present time, for example, lasers that bounce off satellites can be used to measure the movement of the tectonic plates and the vibrations of the ground around volcanoes.[80]

Moreover, we have also seen that when trying to understand our own planet, we often are unaware of many of the crucial factors that affect it. That is so because those factors often are the result of trends and forces that have developed sometimes for billions of years or because they are hidden from our view or caused by interactions with the rest of the solar system. So we make up theories to guide our thinking. But those theories have little contact with the results that might shape them in fruitful directions if we limit ourselves to direct observations of the Earth. Two generations ago we had observed up close only one planet: the Earth. Today we have examined over forty planetary bodies in the solar system. Those other bodies give us the opportunity to test the mettle of our ideas about our own world. The inevitable transformation of those ideas is therefore a transformation of our understanding of the Earth.

A striking illustration of using the solar system as a laboratory is the proposal to terraform Mars, presuming that we find no native life there. Once terraformed, Mars would have an atmosphere similar to Earth's, protected from the solar wind by an artificial magnetic field. It would have oceans, lakes, and rivers. Much terrestrial vegetation would thrive on it, as would many animals and bacteria. The purpose is to make Mars home to large human colonies, perhaps in the hundreds of millions of inhabitants. According to NASA scientist Christopher McKay, using nontoxic super-greenhouse gases such as perfluorinated compounds, we could give Mars a warm climate (15 degrees Celsius average) within one hundred years.[81] This would give us time to work on the atmosphere and other aspects of the Martian environment, which probably would have to go through several stages, depending on the actual plans we decide to follow. At some points we may need to bring in materials from the asteroid belt or even further away. All in all, it would take a few centuries. Carrying out this extraordinary experiment would give us extremely insightful views of what a planetary environment is and how its components evolve and interact with one another. The terraforming of Venus has also been proposed.[82]

We are still rather in the position of a blind man in a china shop. If he moves, he loses; if he does not move, he loses, too. Neither recklessness nor paralysis is to be recommended. To avoid them both, he needs to know what the shop is like and how he can move about in it. To give him sight would be the greatest gift. The Earth is our china shop, and the satisfaction of our curiosity through space science can help us see where we are going. Wisdom requires that we accept that gift.

Notes

1. C. Sagan, *Pale Blue Dot* (Random House, 1994), p. 222. I have adapted this section so far from Sagan's book.
2. Several theories would posit a cold accretion of the Earth. If the Earth had condensed from a gas nebula, a cold accretion would make sense. But on the prevailing view that the Earth accreted from planetesimals, I do not think it makes as much sense. Recent calculations do suppose a cold accretion—the reasoning is that when planetesimals come together at low relative velocities, they can easily stick together; this result allows for the expected quick planetary accretion. At low relative velocities, however, not much heat is produced. At high relative velocities, on the other hand, the colliding planetesimals would vaporize; and thus accretion would take a long time

(see Alan P. Boss, "The Origin of the Moon," *Science* 231 [January 24, 1986]: 341–345). Nevertheless, it seems to me, that the quick initial accretion of bodies with low relative velocities leading up to the proto-Earth would be eventually followed by collisions at high relative velocities with bodies of varying sizes, long before the completion of the Earth's accretion. As long as the proto-Earth's mass was significantly higher than those of the other objects, some of the mass of those objects would be quickly accreted into the Earth by gravitational attraction, even if those objects vaporized upon impact. This "hot" phase in the formation of the Earth is quite reasonable in view of the presently favored hypothesis of the origin of the Moon, which would have the Earth colliding with a body the size of Mars (see later). Surely, if this is plausible, we should favor a scenario in which the proto-Earth is constantly bombarded by bodies too small to break it up but large enough to provide for a "hot" phase during its accretion.

3. Seventy percent of the heat from the interior is dissipated in the motions of the plates. M. Carr, R. S. Saunders, R. G. Strom, and D. E. Wihelms, *The Geology of the Terrestrial Planets* (NASA SP-469, 1984), p. 76.

4. For an account, see S. Schneider, *Laboratory Earth* (Basic Books, 1997), pp. 225–229. Since presumably life could not have survived under the corresponding lower temperatures, several writers have proposed a variety of mechanisms. C. Sagan and G. Mullen first suggested a large greenhouse effect driven by ammonia and methane. Then T. Owen and others argued that large concentrations of CO_2 were more likely than ammonia (up to 1,000 times today's CO_2 levels). Most hypotheses depend on a large greenhouse effect created by the large outgassing from the interior of a hot young planet.

5. Schneider, *Laboratory Earth*, p. 261.

6. Schneider, *Laboratory Earth*, p. 261.

7. For a report, see R. A. Kerr, "Milankovitch Climate Cycles through the Ages," *Science* 235 (February 27, 1987): 973–974.

8. O. B. Toon, J. B. Pollack, and K. Rages, "A Brief Review of the Evidence for Solar Variability on the Planets," in *Proceedings of the Conference on the Ancient Sun*, ed. R. O. Peppin, J. A. Eddy, and R. B. Merrill (1980), pp. 523–531.

9. L. W. Alvarez, W. Alvarez, F. Asaro, and H. V. Michel, "Extraterrestrial Cause for the Cretaceous-Tertiary Extinction, *Science* 208 (1980): 1095–1108.

10. Some interesting studies by D. M. Raup and J. Sepkoski suggested that the extinction of a significant portion of terrestrial life is a periodic occurrence, with the period being about twenty-six million years (D. M. Raup and J. J. Sepkoski, Jr., "Mass Extinctions in the Marine Fossil Record," *Science* 215 [1982]: 1501–1503. For several of the issues raised in the last few paragraphs, the reader may wish to consult Ch. VI of *The Evolution of Complex and Higher Organisms*, ed. D. Milne, D. Raup, J. Billingham, K. Niklaus, and K. Padian (NASA SP-478, 1985). According to a hypothesis by Raup, this extinction rate depends on the orbit of a star companion to the Sun, a dwarf star dubbed "Nemesis" which causes the gravitational disturbances described in the text. Nemesis was never found, though. (For a very accessible account, read D. M. Raup, *The Nemesis Affair* [W.W. Norton & Co., 1986]). This idea has fallen out of favor, since Nemesis was never found.

11. See D. Desonie, *Cosmic Collisions*, a Scientific American Focus Book (Henry Holt & Co., 1996).

12. If they are not, we will still come away with sharpened alternative accounts of the fate of living things.

13. These estimates come from Desoinie's *Cosmic Collisions*, pp. 100–101.

14. Since the Earth is a bit more massive than Venus, its gravitational attraction is consequently larger. On the other hand, Venus is closer to the Sun, and thus objects with pronounced elliptical orbits and rather small perihelions are bound to pass closer to Venus than to Earth. It is unfortunate, for the purposes of statistical prediction, that smaller craters than Meteor Crater (whose creation would be disastrous enough) do not register on the surface of Venus—such meteors burn up in the extremely dense atmosphere of that planet. The present estimates for objects around 60 meters in diameter strike me as being at least of the right order of magnitude and probably quite accurate. Incidentally, my own estimates involve some circularity, since the age of the surface of Venus has been estimated using the rate of cratering (although such rate has been calibrated to some degree with actual measurements on the Moon).

15. Thermonuclear weapons are the first choice, although O'Neill's mass drivers might also do the job. He envisioned using such drivers to transport asteroids rich in valuable minerals to a lunar orbit. The effectiveness of nuclear bombs is the subject of some controversy and the inspiration for several movies.

16. For an alternative account of the extinction of the dinosaurs, see R. T. Bakker, *The Dinosaur Heresies* (William Morrow and Co., 1986). The Alvarez account has become the received view, however.

17. To Darwin's theory, also called the fission theory, Osmond Fisher added the hypothesis that the Moon had come out of what is now the Pacific Ocean basin. In this form the theory was popularized in the first decades of twentieth century. For an account, see S. G. Brush, "Early History of Selenogony," in *Origin of the Moon*, ed. Hartmann et al. (Houston, 1986), pp. 3–15.

18. Brush, "Early History of Selenogony."

19. Brush, "Early History of Selenogony." See also Brush, "Harold Urey and the Origin of the Moon: The Interaction of Science and the Apollo Program," in the *Proceedings of the Twentieth Goddard Memorial Symposium* (1982), published by the American Astronautical Society; and "From Bump to Clump: Theories of the Origin of the Solar System 1900–1960," in *Space Science Comes of Age: Perspectives in the History of the Space Sciences*, ed. P. A. Hanle and V. D. Chamberlain (Smithsonian Institution Press, 1981), pp. 78–100.

20. Boss, "The Origin of the Moon."

21. Boss, "The Origin of the Moon." Another important piece of evidence is the fact that the Moon seems to have a very small core, about 300 to 425 kilometers in radius, holding about 4% of the Moon's mass. Had the Moon been born side by side with the Earth, or had it been captured, it should be expected to have a much more significant core. *Science News* 155 (March 27, 1999), p. 198.

22. Junjun Zhang, Nicolas Dauphas, Andrew M. Davis, Ingo Leya, and Alexei Fedkin, "The Proto-Earth as a Significant Source of Lunar Material," *Nature Geoscience* 5 (March 25, 2012): 251–255.

23. See S. G. Brush, "Nickel for Your Thoughts: Urey and the Origin of the Moon," Science 217 (September 3, 1982): 891–898. Maria could have also been produced by magma flowing from hot zones of convection cells. See P. Cassen, R. T. Reynolds, F. Graziani, A. Summers, J. McNellis, L. Blalock, "Convection and Lunar Thermal History," in *Solid Convection in the Terrestrial Planets, Physics of the Earth and Planetary Interiors*, ed. P. Cassen, vol. 19 (1979), pp. 183–196. A radically different alternative, according to which dust carried by low electrical currents created the maria, was suggested by T. Gold. It is described by B. W. Jones in *The Solar System* (Pergamon Press, 1984), pp. 177–179.

24. Boss, "The Origin of the Moon."

25. See Jones, *The Solar* System, p. 183 and pp. 203–207.

26. I will discuss another significant contribution in Chapter 5.

27. According to Jones, the oldest rock found on the Moon is 4.6 billion years old (a silicate of a type called dunite). Jones, *The Solar* System, p. 173.

28. Collisions with asteroids may have also broken open lakes of molten material near the crust, and that material might have then spilled over to form the maria. The molten material could have resulted from the heat of radioactive elements or even from previous collisions with giant asteroids.

29. A significant difference, however, may be that Ganymede is about 50% water, and so its crustal movements are closer to ice tectonics than those on Earth.

30. The Tsar Bomba, the most powerful nuclear device ever tested, yielded around 50 megatons. S. Narayanan, *The Tsar Bomba*, http://large.stanford.edu/courses/2015/ph241/narayanan2/

31. By mechanisms to be explained later, most of the carbon also ended up in the atmosphere, which explains the extraordinary amount of carbon dioxide.

32. Carr et al., *The Geology of the Terrestrial Planets*, p. 77.

33. There are some who still manage to see something resembling the Earth's mid-ocean ridges, but even if we grant that, it still seems a far shot from the full-blown terrestrial plate tectonics.

34. D. H. Grinspoon, *Venus Revealed* (Addison-Wesley, 1997), p. 179.

35. One might think that the representativeness of the sample is justified by the very consideration that otherwise it would be too big of a coincidence. But this idea is questionable, as we will see later in the text.

36. J. Leslie, *The End of the World* (Routledge, 1996). Leslie believes that this statistical reasoning should make us more fearful of possible cosmic cataclysms, such as giant asteroid impacts and space-time-gobbling new universes growing inside our universe, as well as human-made catastrophes.

37. S. E. Lauro, E. Pettinelli, G. Caprarelli, L. Guallini, A. P. Rossi, E. Mattei, B. Cosciotti, A. Cicchetti, F. Soldovieri, M. Cartacci, F. Di Paolo, R. Noschese, R. Orosei, "Multiple Subglacial Water Bodies below the South Pole of Mars Unveiled by New MARSIS Data," *Nature Astronomy* 5(2021): 63–70. https://doi.org/10.1038/s41550-020-1200-6

38. P. R. Christensen, "Water at the Poles and in Permafrost Regions of Mars," *Elements* 3, no. 2 (2006): 151–155, doi:10.2113/gselements.2.3.151.

39. Some researchers have suggested that the Earth's early atmosphere, following the initial heavy bombardment by asteroids, also had a very high percentage of CO_2. The ensuing greenhouse would then compensate for the dimmer Sun. Life eventually removed much of the CO_2, thus preventing a runaway temperature when the Sun's luminosity increased. This hypothesis runs contrary to other ideas on the composition of the early atmosphere, according to which a primitive atmosphere would exhibit either a highly reducing mixture of methane, ammonia, water, and molecular hydrogen (similar to that of Jupiter, Saturn, and the other planetary gas giants) or else a mildly reducing mixture of carbon monoxide, carbon dioxide, nitrogen, and water, with not much molecular hydrogen. (In this context a mixture is reducing to the extent that it contains hydrogen.) To decide between these and perhaps other alternatives, it will be helpful to study not only the histories of Mars and Venus, but the largely methane atmospheres of Titan and Triton, the large moons of Saturn and Neptune, respectively. The reason these matters are so worth looking into is that knowing more about the composition of the primitive atmosphere can tell us much about the origin and evolution of the global environment of a planet; in this case, of our planet.

Consider also one of the most interesting aspects of Jupiter's atmosphere: the famous Red Spot. In a dense atmosphere with winds of hundreds of miles per hour, how could a storm, which is what the Red Spot is, remain stable for centuries, perhaps for many thousands of years? The answer seems to be that the fast spin of Jupiter (once every ten hours) produces very strong Coriolis forces, which in turn produce the turbulent winds that drive the gigantic eddy of gas otherwise known as the Red Spot. As the planet spins, many smaller eddies develop, but these eddies eventually feed the Red Spot. In the midst of turbulence the Red Spot has achieved stability within the Jovian atmosphere. Thus stability arises from chaos. But interesting as this may be, what significance does it have for people on Earth? The significance is that space scientists see many parallels between the dynamics of the Red Spot and some weather patterns in the atmosphere of the Earth. In particular, these scientists see parallels to systems of high pressure that sit still for weeks or even months. Understanding this phenomenon, known as "blocking," would be a great help in forecasting the weather here on Earth. Of course, it may still turn out that what we learn about the stability of the Red Spot does not apply to stationary high-pressure systems on the Earth.

40. A. Yin, "Structural Analysis of the Valles Marineris Fault Zone: Possible Evidence for Large-Scale Strike-Slip Faulting on Mars," *Lithosphere* 4, no. 4 (2012): 286–330.

41. They have that name because they work in the manner of a greenhouse: they let the sunlight through but trap the heat in.

42. For an account, see H. D. Holland, B. Lazar, and M. McCaffrey, "Evolution of the Atmosphere and Oceans," *Nature* 320 (March 6, 1986): p. 33. For variations in CO_2 levels throughout the history of the planet, see E. T. Sundquist and W. S. Broecker, *The Carbon Cycle and Atmospheric CO2: Natural Variations Archean to Present* (American Geophysical Union, 1985).

43. Recently some researches have argued that on the basis of such evidence we cannot prove that the global temperature was higher then.

44. Life in the Universe. Proceedings of the Conference on Life in the Universe, held at NASA Ames Research Center, June 19–20, 1979. Editor, John Billingham; Publisher, MIT Press, Cambridge, Massachusetts, 1981. ISBN # 0-262-52062-1. LC # QB54 .L483, P. 79, 1981.

45. P. G. Falkowski, "The Ocean's Invisible Forest," *Scientific American* 287, no. 2 (August 2002): 56–57.

46. Peter D. Ward and Donald Brownlee, *The Life and Death of Planet Earth* (Times Books, 2002), p. 61.

47. Ward and Brownlee, *The Life and Death of Planet Earth*, p. 57.

48. Adapted from L. Margulis and D. Sagan, *Microcosmos* (Summit Books, 1986), p. 237.

49. Lovelock and Margulis, "Atmospheres and Evolution," p. 81.

50. This is a recurring theme in Lovelock's work. See his article in "A Physical Basis for Life Detection Experiments," *Nature*, London, No. 207, p. 568.

51. See Lovelock's analysis of J. and P. Connes (*Journal of the Optical Society of America* 9 (1966): p. 896, in Lovelock and A. J. Watson, "The Regulation of Carbon Dioxide and Climate: Gaia or Geochemistry," *Planetary and Space Science* 30, no. 8 (1966): p. 795.

52. Ward and Brownlee, *The Life and Death of Planet Earth*, Mcmillan (2004) pp. 71–86.

53. R. Chris Wilson, Stephen Drury, and Jenny L. Chapman, *The Great Ice Age* (2000). Cited by Ward and Brownlee, *The Life and Death of Planet Earth*, p. 82.

54. Ward and Brownlee, *The Life and Death of Planet Earth*, p. 83. On the other hand, global warming, according to some speculations, might trigger the ice age by shutting off the water circulation patterns in the Atlantic (warm water going north closer to the surface and cold water going south closer to the bottom of the ocean).

55. J. E. Lovelock and M. Whitfield, "Life Span of the Biosphere," *Nature* 296 (1982): 561–563.

56. This account is borrowed from Ward and Brownlee, *The Life and Death of Planet Earth*, pp. 101–116.

57. Life managed to survive the opposite kind of hell: an Earth completely covered by ice. Ward and Brownlee, *The Life and Death of Planet Earth*, p. 75.

58. Ward and Brownlee, *The Life and Death of Planet Earth*, p. 50. See also Schneider, *Laboratory Earth*, p. 66.

59. L. K. Fenton, P. E. Geissler, & R. M. Haberle, "Global Warming and Climate Forcing by Recent Albedo Changes on Mars," *Letters to Nature* 446(April 5, 2007), doi:10.1038/nature05713.

60. N. Shaviv and J. Veizer, "Celestial Driver of Phanrozoic Climate?" *Geological Society of America* 13 (2003): 4–10.

61. H. Svensmark, "Influence of Cosmic Rays on Earth's Climate," Danish Meteorological Institute, *Physical Review Letters* 81 (1998): 5027–5030.

62. H. Svensmark, "Cosmoclimatology: A New Theory Emerges," *Astronomy and Geophysics* 48 (2007): 1.18–1.24.

63. Svensmark, J., M. B. Enghoff, N. Shaviv, and H. Svensmark, "The Response of Clouds and Aerosols to Cosmic Ray Decreases," *Journal of Geophysics and Research in Space Physics* 121 (2016): 8152–8181, doi:10.1002/2016JA022689.

64. Henrik Svensmark and Juliane M. Vej, "Evidence of Nearby Supernovae Affecting Life on Earth," *Monthly Notices of the Royal Astronomical Society* 423 (2012): 1234–1253.

65. According to Benford, it would take a land the size of Australia covered with trees to soak up all the present increase in CO_2 "Climate Controls," *Reason* 29, no. 6 (November 1997): 24–31.

66. For a very interesting and readable account, see Jonathan Lambert, "The Forested Farms of the Future," *Science News*, July 3, 2021 & July 17, 2021, pp. 30–35.

67. Jennifer Couzin, "Landscape Changes Make Regional Climate Run Hot and Cold," *Science* 283 (January 15, 1999): 317–319.

68. Couzin, "Landscape Changes Make Regional Climate Run Hot and Cold," p. 27.

69. McKibben, *The End of Nature*.

70. McKibben, *The End of Nature*.

71. Falkowski, "The Ocean's Invisible Forest," p. 61.

72. Andrew Lawler, "Stormy Forecast for Climate Science," *Science* 305 (August 20, 2004): 1094–1097. Unfortunately, I fear, the needed effort may be undermined by the political advocacy by scientists during the past several years. According to this article, "Whereas fiscal conservatives would attack any massive new research program as unaffordable, liberals are likely to see it as a ruse to delay action on the underlying problems that are causing global warming" (p. 1097).

73. D. Goto, H. Yoshida, H. Suzuki, K. Kisara, and K. Ohashi, *The Overview of JAXA Laser Energy Transmission R&D Activities and the Orbital Experiments Concept on ISS-JEM*, Proceedings of the International Conference on Space Optical Systems and Applications (ICSOS), S5-2, Kobe, Japan, May 7–9, 2014.

74. J. C. Mankins, *SPS-ALPHA: The First Practical Solar Power Satellite via Arbitrarily Large Phased Array (A2011-2012NASANIACPhase1Project)*, Final Report to NASA and NIAC (2012), http://www.nss.org/settlement/ssp/library/SPS_Alpha_2 012_Mankins.pdf

75. J. T. Early, "Space-Based Solar Shield to Offset Greenhouse Effect," *Journal of the British Interplanetary Society* 42 (1989): 567–569.

76. For updated information I have been helped by Kenneth Roy's draft "The Solar Shield Concept: Current Status and Future Possibilities," under consideration for publication.

77. M. C. Evans and C. S. Ruf, "Toward the Detection and Imaging of Ocean Microplastics with a Spaceborne Radar," *IEEE Transactions on Geoscience and Remote Sensing*, doi:10.1109/TGRS.2021.3081691.

78. Ian Crawford, "Lunar Resources: A Review," *Progress in Physical Geography* 39 (2015): 137–167.

79. James S. J. Schwartz, *The Value of Science in Space Exploration* (Oxford University Press, 2020).

80. One interesting illustration of this point can be found, once again, in the environmental problem of CO_2. Most of the dire scenarios include the melting of the polar caps. But determining how the caps would melt needs at least two kinds of investigation: (1) a way of measuring changes in the ice cap that can be correlated with increases in global temperature, and (2) a general theory about the formation and evolution of ice caps that will allow us to infer trends from such measurements. Fortunately space exploration has given us the means to take accurate measurements

that would be practically impossible otherwise: polar satellites that make the precise comparisons needed for a fine determination of changes in the ice caps.

81. Christopher P. McKay, "Make Mars Great Again: How to Terraform a Room-Temperature Mars in 100 Years," *Nautilus*, December 15, 2016.
82. Kenneth Roy, "Terraforming Venus, and Similar Planets, Using a Pneumatically Supported Shell," Interstellar Research Group Meeting, Wichita, Kansas, 1989.

5

Cosmology and Fundamental Physics

According to the journal *Science*, Rashid Sunyaev, a famous Russian astrophysicist, "once heard the chair of his department say that 'astronomy was an absolutely useless science.'"[1]

After the spectacular successes of the space telescopes and the new generation of Earth-bound telescopes, the public may be surprised to learn that not long ago many scientists regarded space astronomy and space physics with some suspicion. Quite a few physicists, for example, felt that all those billions of dollars for space astronomy should have supported the construction of a new generation of particle accelerators instead—particle accelerators dealt with truly basic science. I presume that a good many of those physicists may now agree that the money spent in space telescopes has been money well spent. But it is important to see *why* they are right in having changed their minds.

The reason is that, in the pursuit of cosmological knowledge, physics and astronomy done in space affect the transformation of our views in a very important respect: they provide a framework within which to challenge our most fundamental terrestrial sciences. For to understand the formation and evolution of the universe, we need to see how the basic laws of nature are expressed in it. At the same time, to have a good grasp of the basic laws of nature, we need to see how well we can describe the universe by using them. In a second respect they affect that transformation in an even more radical way: astronomy and physics done in space allow us to discover phenomena that we could not have discovered otherwise and that will force us to develop a new physics.

The Objection to Space Science

As we recall from Chapter 3, a scientific critic might argue that, since these sciences examine faraway objects, the ensuing transformation of our views

The Dimming of Starlight. Gonzalo Munévar, Oxford University Press. © Oxford University Press 2023.
DOI: 10.1093/oso/9780197689912.003.0005

is unlikely to pay off for the inhabitants of the Earth. For example, space astronomy and physics constitute a prime example of attempting to satisfy our intellectual curiosity—they aim to describe features of the universe that many people find interesting, sometimes fascinating. The problem for my thesis is precisely that these space sciences fit my points about curiosity so well while *apparently* failing to satisfy my expectation about practical results in the long run. Surely, the objection continues, it is by no means obvious that the transformation of our theories about black holes, quasars, and intergalactic gas will be of much application on the Earth.

There is, according to this objection, a great difference between Earth-bound physics and these space sciences. Consider anew the example of how Einstein's revolution in physics led to lasers and their application in medicine: that revolution transformed our understanding of the basic principles of matter, of principles that apply *down here*. It is not surprising, then, that our panorama of problems and opportunities was bound to change as a result of the transformation of our thought. The principles of fundamental science (e.g., relativity or particle physics) apply down here because they apply everywhere. By contrast, space astronomy and space physics merely apply to stars, galaxies, and quasars, the fundamental principles of matter discovered by Earth-bound physics—thus, they are derivative sciences. My thesis about serendipity would then apply only to fundamental science. Therefore, space astronomy and space physics cannot be justified by my general philosophical argument.

Let me begin the discussion of this objection with a counterexample. Astrophysicists Anil Pradhan and Sultana Nahar of Ohio State University were investigating the composition of stars by analyzing the flow of radiation through them. They became particularly interested in how iron and other heavy metals release a flood of low-energy electrons after absorbing X-rays in a narrow range of frequencies. The bombardment of iron by X-rays can be observed in the vicinity of black holes as well. These astrophysicists realized that their discovery could be used to treat cancer: nanoparticles made of heavy metals could be introduced into the tumors and then subjected to low doses of X-ray radiation. The resulting flood of electrons would destroy the tumor without causing much damage to healthy tissue. According to Pradhan, this is the most fundamental advance in X-ray production since 1890.[2]

For almost a century now, the most fundamental and empirically successful description of matter has been given by the so-called Standard Model, which explains the universe in terms of its building blocks (particles) and the fundamental forces (strong, weak, electromagnetic, and gravitational) that allow those blocks to interact. The main experimental tools of the Standard Model have been giant particle accelerators that smash those particles at speeds close to that of light. Physicists then theorize from the resulting debris. As I mentioned earlier, when the choice came between spending billions to build even more powerful accelerators or spending billions to put up telescopes in orbit, the feeling among many physicists was that interesting though astronomy may be, taking money away from the terrestrial tools that would allow us to advance the Standard Model further was tantamount to blunting fundamental science's cutting edge. That feeling is gone today, even though the proponents of the Standard Model are thrilled with the reported discovery of the Higgs particle, which would presumably explain how matter acquires mass.[3] This explanation would, of course, be greatly relevant to the exploration of the universe.

My general response will proceed as follows: (1) space physics and astronomy have distinct scientific advantages over terrestrially bound sciences; (2) these scientific advantages show that space physics and astronomy are fundamental sciences in the same sense that terrestrially bound physics is, because you cannot do terrestrial physics properly without doing space science.

New Physics: The Dark Side

It had been suspected for some time that galaxies had far more matter than we could determine from what was visible. Once we were able to look at galaxies in the full spectrum, we realized that perhaps as much as 90% of some galaxies' mass was unaccounted for (one main reason for this realization is that the outer stars in a galaxy are going so fast that without the gravitation of that much extra mass to keep them in, they would be flung into intergalactic space). But apparently most of that missing mass (now called "dark matter") cannot be seen, for it does not interact with ordinary matter through any of the fundamental forces except gravity. Moreover, the universe contains about five times more dark matter than regular (Standard Model) matter. In other words, most of the matter in the universe is completely different from anything we have known until now (it is not made up of protons, neutrons, electrons, and the like). To study dark matter, it is necessary to observe the whole universe in a great variety of ways, including space telescopes. The new fundamental physics, thus, requires cosmology.

To make matters worse, it was discovered that the expansion of the universe is accelerating, in violation of the sensible belief that gravitational attraction should slow down and perhaps even reverse the expansion initiated in the famous Big Bang. A new form of energy, dark energy, which we understand even less than we understand dark matter, presumably accounts for that perplexing expansion (see Figure 5.1).[4]

Since dark energy (68%) and dark matter (27%) make up most of the matter-energy of the universe, we find that the Standard Model tries to explain the entire universe on the basis of the 5% of matter-energy that it is acquainted with. Imagine that you come to a new place and get to know only 5% of it. All you know about the rest is that it is vastly different from the small portion you know. How confident would you feel about explaining the whole on the basis of the one part you can handle?

To study dark energy, we are advised to go into space, sooner or later. Some of the work of surveying the universe can be done with terrestrial telescopes, but the findings of such surveys will have to be corroborated and

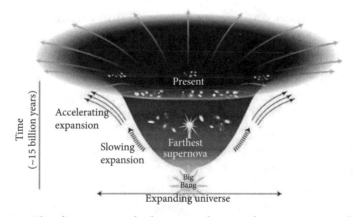

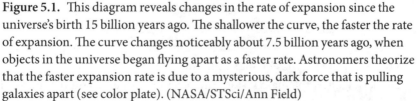

Figure 5.1. This diagram reveals changes in the rate of expansion since the universe's birth 15 billion years ago. The shallower the curve, the faster the rate of expansion. The curve changes noticeably about 7.5 billion years ago, when objects in the universe began flying apart as a faster rate. Astronomers theorize that the faster expansion rate is due to a mysterious, dark force that is pulling galaxies apart (see color plate). (NASA/STSci/Ann Field)

supplemented by telescopes in orbit, as we will see later. This means that to have much of a chance to come up with a fundamental explanation of the universe we need to do space science.

Some physicists still hope that in the new particle accelerators, which will produce very high energies, violent collisions will yield some dark matter particles—so-called WIMPS and axions are the main candidates. And, who knows, their hopes may perchance be realized, but since we do not know what dark matter is, those physicists sound a bit too optimistic. And even then, we would still have the even bigger puzzle of dark energy.

To do away with the problem of dark matter, some physicists have proposed a variety of modified theories of gravity. These hypotheses, which are rejected by most physicists, would, of course, radically transform fundamental physics, a transformation motivated by space astronomy and physics. Whether the consensus view or an alternative wins the day, space science should receive significant credit for the serendipity that will result from the soon-to-be new physics.

In the rest of the chapter, I will place the discussion in a historical context, and then I will offer a variety of examples to illustrate some of the many ways in which space science will help guide fundamental physics.

Space Science and the Tradition of Fundamental Physics

Let us begin our discussion of the advantages of space physics and astronomy over terrestrially bound sciences by remembering that the connection with astrophysics has been a trademark of modern physical science from its inception and throughout most of its history.

Although it is well known that Copernicus placed the Sun and not the Earth at the center of the universe, his motivation is not so well understood. It was not that his system could account for the position of the planets clearly better than the Ptolemaic system, for even Copernicus acknowledged that the matter was not settled. Nor was it obvious either, in spite of the claims by Pierre Duhem and others, that his system was vastly simpler. It is true that the Ptolemaic system employed a variety of mechanisms—epicycles, eccentrics, deferents, equants—to account for the paths of the planets, but with the exception of the equant so did the Copernican system.[5] The difference was that the Ptolemaic system often had alternative combinations of such mechanisms for different aspects of the behavior of the same planet. This would seem outrageous to someone weaned on the notion that *only one such mechanism* could be correct. But the mere talk of correctness assumes that we can inquire about the real nature of the heavens.

We feel entitled to make that assumption rather freely today. But that was not the case in Copernicus's day. From the time of Ptolemy (second century CE), the inquiry about the reality of the heavens had been looked upon with suspicion. The reason was that whereas the progress of mathematical astronomy made it possible to calculate with increasing precision the positions of the planets, the accounts of *why* the planets moved as they did had broken down not long after Aristotle had proposed the interaction of concentric spheres made of his quintessence (about 350 BCE).[6]

According to Ptolemy himself, mathematics can apply only to "changes in form: i.e. in trajectory, shape, quantity, size, position, time, and the rest."[7] As to the actual nature of things there is little that science can do because they either take "place far above us, among the highest things in the universe, far away from the objects we directly observe with our senses," or else, as the objects of (terrestrial) physics, those "material things . . . are so unstable and difficult to fathom that one can never hope to get philosophers to agree about them."[8] Questions about the nature of the heavenly objects must lead one back to the ultimate source of all change, and thus they can only be answered by theology. Therefore, science gains little profit from asking them. And they are also distinct in kind from the sorts of questions that physics tries to answer, whose underlying principles, if any, did not seem amenable to mathematical treatment.

In the long run, the Copernican Revolution accomplished several important changes in points of view. For one thing it insisted on investigating the real nature of the behavior of the heavenly objects. And it did so by looking for mutual underlying mathematical principles for both the heavens and the Earth. The success of this gross violation of Ptolemy's methodological rules turned on Copernicus's belief that astrophysics was possible. Eventually Newton succeeded where Aristotle had failed, and astrophysics became the shining example that new branches of physics had to follow.

Confusions about Copernicus's motivation were created mainly by Osiander's preface to Copernicus's masterpiece, *On the Revolutions*. Fearing a confrontation with theological dogma, Osiander urged the readers to "permit these new hypotheses to become known together with the ancient hypotheses," and to do so because Copernicus's hypotheses are "admirable and also simple and bring with them a huge treasure of skillful observations." But Copernicus's readers, Osiander wrote, should not accept as the truth "ideas conceived for another purpose."[9] All these admonitions by Osiander contradicted Copernicus's own words and belied his attempt to discover the truth about the heavens by rational means instead of revelation.

In the centuries following Copernicus, astrophysics continued as a driving force of fundamental theory. Newton is, of course, the most prominent example. His laws of dynamics applied equally to terrestrial and heavenly objects, and his law of gravitation was a striking statement of the discovery that the force that kept us glued to the surface of the Earth was the same that made the stars and planets keep their appointed rounds.

But you need more than great theorists. Fundamental questions often cannot be asked without the appropriate technology and will not be asked without the right kind of inspiration and motivation. This realization leads to another: a whole host of activities are potentially as crucial to scientific progress as work that aims to solve problems within the most prestigious field of the time. Researchers who create new technology or new opportunities may contribute just as much to keep intact the dynamic character of science. And inspiration has often come as much from the planets as from the stars.

In the case of planetary science, as Stephen Brush reminded us in 1979, the attempt to satisfy intellectual curiosity brought about many of the most significant advances of the past few centuries. In 1746, for example, d'Alembert won a prize for making one of the first consistent uses of partial differential equations. His was an essay on winds. The work of Laplace at the end of the eighteenth century resulted from the calculations in celestial mechanics of Clairaut, Euler, and Lagrange. Legendre and Laplace originated the use of spherical harmonics in potential theory while trying to calculate the gravitational attraction of the Earth. This use was later of great value in electricity and in quantum mechanics. And many of Poincare's major works on mathematical analysis were inspired by problems in planetary mechanics.[10]

Similar remarks may be made about Gauss, one of the world's greatest mathematicians, who did much work in geodetic surveys and terrestrial magnetism. As Brush points out, "even an advocate of pure science might concede that Gauss' geophysical work provided the stimulus for some of his contributions to geometry and potential theory, just as his early work on the computation of orbits led to a major contribution to probability theory, the 'Gaussian distribution,' and could thereby be justified."[11] James Clerk Maxwell, clearly a giant of physics, won the Adams Prize with an essay on the stability of the rings of Saturn. This work was the basis for his kinetic theory of gas viscosity and eventually of his theory of transport processes. He later returned to the problem of the rings of Saturn and applied to it the methods he developed in his kinetic theory of gases.

A case of particular interest to Brush in establishing the connection between pure science and planetary science is the nineteenth-century problem of the dependence of thermal radiation on the temperature of the source. This problem was of great significance to planetary science, of course, because the Sun is the greatest source of radiation in the solar system. The outcome of the attempt to determine the surface temperature of the Sun "was Stefan's suggestion (1879) that the data could best be represented by assuming that the rate of emission of energy is proportional to the 4th power of the absolute temperature."[12] Further experimental investigation, and Boltzmann's theoretical derivation of such a formula, led to the subsequent work on the frequency distribution of black-body radiation. The search for the law that would govern such distribution can thus be seen as the genesis of Planck's quantum theory.

These remarks on the history of physics vindicate the claim that the study of the heavens at all levels has been a driving force in the development of fundamental science. This may come as a surprise to some. But it should not be a surprise if we consider the variety of interactions between cosmology and other areas of science, and between the different levels of cosmological research. The study of the cosmos leads to the discovery of fundamental principles of physics; further development of physics in turn leads to new investigative tools of the cosmos; and so on. We may speak here of a dialectical relationship between areas and levels of science which results in the dynamic growth of science. In the case of planetary science, we find no exception: fundamental research on physics and cosmology leads to changes in our ideas about the solar system and its planets. On the other hand, those ideas in turn give many hints as to useful lines of "pure" inquiry. This result is not merely part of the historical record: if anything, it should receive greater prominence as space exploration multiplies our means of investigating the cosmos.

Indeed, we will see presently that space science provides an excellent opportunity to enhance the relationship between astrophysics and the rest of fundamental physics.

A New Vision of the Universe

When Galileo turned his telescope to the heavens, a new sense was born. Galileo's telescope gave us the moons of Jupiter, the phases of Venus, and thousands of new stars. But this new sense was no mere addition, for it helped usher in a view that contradicted direct sensory experience. As Galileo himself said, "there is no limit to my astonishment, when I reflect that Aristarchus and Copernicus were able to make reason so to conquer sense that, in defiance of the latter, the former became mistress of their belief."[13] But in the telescope he found a "superior and better sense than natural and common sense."[14] This new sense, furthermore, was not just a refinement of sight, but rather an alternative that agreed with the Copernican view, unlike plain sight. Indeed, whereas to plain sight the magnitude of Mars changed little—a troublesome fact for the Copernican view according to which the distance between Mars and the Earth varied considerably—in looking through the telescope the magnitude behaved as if God had been a Copernican. Likewise, the phases of Venus explained why the magnitude of Venus remained constant: when Venus is closest, we only see a small portion of its disk lighted, but as it moves away from us, we see more and more of its disk lighted until, when it is furthest from Earth, we see it fully lighted.

In this way a new technology came to the rescue of an idea that was to transform our view of nature most profoundly. Not that the telescope was free from reasonable question, for on the contrary, given what was known about optics and perception at the time, its reliability was a lucky assumption made plausible more by Galileo's enthusiasm than by his argument. Among other things, we should keep in mind that the brain does not merely experience the pattern of light that strikes the retina: it interprets that pattern on the basis of past experience, expectation, and environmental clues. It also imposes constancies of shape, color, and size upon the visual image. When peering into the skies, however, many of those clues are absent and thus the untrained eye can experience many illusions. And indeed, many people looking through Galileo's telescope did not see what he saw.[15] Moreover, it was generally thought that perception cannot be trusted when the natural medium through which the information travels has been tampered with (e.g., a haze or a drunken brain). Galileo's telescope was theoretically suspect, then, because it clearly altered the natural medium through which the light from distant objects was transmitted to the eye.

But the telescope did open up many avenues of observation and investigation that would not have been there otherwise. It was a promise from the heavens better kept in the course of the new science than perhaps Galileo had a right to imagine. For him it was too striking a coincidence that the telescope would so match the new astronomy of Copernicus. For others—whose fundamental views were at stake—it was a case of a distorting instrument of observation presuming to support a refuted and obnoxious view of the cosmos.

To most of us now, it seems fortunate that, as Galileo put it, Copernicus "with reason [theory] as his guide . . . resolutely continued to affirm what sensible experience seemed to contradict."[16] Many of those who were in a position to choose decided in favor of the most exciting of the alternatives (in accordance with a principle that NASA scientist Brian Toon fondly calls "Sagan's Razor"). Whether they had other, more compelling reasons I shall not discuss here. Suffice it to say that the invention by Newton of the reflecting telescope and the refinement of both kinds of telescopes permitted not only the discovery of many new objects in the universe but also a shift in perspective about its nature. And having embarked on a different approach to nature, the new scientists also had the motivation and opportunity to develop the auxiliary sciences (e.g., optics, electromagnetism) that eventually established the reliability of the telescope as well.

Such reliability is rather limited, as we have learned. Apart from obvious problems such as background lights and bad weather, optical telescopes until recently had to contend with the effects of columns of air within them and of gravitational distortions of their very large mirrors. Worse still, even though a large telescope such as the 5-meter telescope on Mount Palomar could in principle separate the images of objects in the sky as close to each other as 0.02 arc second (this is called angular resolution), the motion of air molecules blurs the path of light through the atmosphere to the point that the best angular resolution we can achieve is about fifty times worse than that. Such was the situation at the point when space technology began to place optical telescopes in orbit, Hubble (1990) and Gaia (2013) being two famous examples. Today those space telescopes work in concert with new generations of Earth-bound telescopes in which computers compensate for some of the distortions of the atmosphere and in which combinations with other such telescopes (in interferometers) allows the collection of much greater amounts of light.

Moreover, we know now things that Galileo could not have the tools to imagine. Light is a form of electromagnetic radiation and the Earth's atmosphere blocks many other forms of that radiation. More specifically, it absorbs the X-ray and gamma frequencies, as well as most of the ultraviolet and several bands of the infrared. The envelope of gases that made life possible in the beginning, and has protected it ever since, has given us only a small window into the universe at large. Through that window we have seen, and we have dreamt, and through that window we have also adjusted our views of nature. Galileo's new sense became better and better, and when it seemed that it was reaching its limitations, in the middle of the twentieth century, we discovered radio astronomy, and then infrared astronomy.

By then, we had found that the Sun was not the center of the universe either, that it was one among billions of stars in the outskirts of a rather common galaxy, and that the galaxies were receding from each other, that is, that the universe was expanding. This last discovery prompted Einstein to recognize what he called his greatest mistake. The idea that the universe might change in size had seemed so preposterous at the time he proposed his general theory of relativity that he introduced a constant into the theory to ensure that the universe would appear static in it.

So much resulted from the evolution of Galileo's instrument. But then, with the discovery of radio and infrared astronomy, new kinds of objects and new kinds of activity in objects already known gave us a glimpse of what a look at the universe in the full electromagnetic spectrum may do. Quasars and pulsars, and the radiation left over from the Big Bang, began to tell us about a universe far stranger than we had imagined. With the advent of space exploration, we can go above the atmosphere to examine the universe in the full electromagnetic spectrum. We can also do far better astronomy in the more traditional wavelengths (radio, optical, and infrared).

For example, in the visible range, NASA's Hubble Space Telescope, unencumbered by most of the handicaps of its terrestrial counterparts, swept a volume 350 times larger and produced images ten times finer than most traditional telescopes could at that time. Thanks to Hubble, the Swift Gamma Ray Burst Explorer, TESS, the Chandra X-ray Observatory, and the Fermi Gamma-ray Space Telescope, as well as other space and ground telescopes, we have confirmed the discovery of 5,171 exoplanets, with 8,933 candidates, in 3,870 planetary systems, as of October 1, 2022.[17]

Radio astronomy, which allows us to study regions obscured by dust, as well as some very cold or very hot objects, is not hampered much by the atmosphere, but it is limited by the size of the Earth. Even though the angular resolution of the largest radio telescopes is very poor (about 10 arc seconds), they can be placed together in large arrays (with the telescopes at large distances from each other) that are equivalent to gigantic telescopes. The Very Long Baseline Array, for example, has a baseline of 8,611 kilometers. Unfortunately, Earth radio signals easily drown out the weak radio signals from space. To help radio astronomy take a leap forward, pun intended, NASA aims to build the Lunar Crater Radio Telescope (LCRT) on the far side of the Moon, where radio signals from Earth are blocked by the Moon itself.[18]

We will see further examples later concerning how not only astrophysics and cosmology but even traditional fundamental physics (the Standard Model plus general relativity) can be advanced by instruments in space. But before that, let us consider a historic-philosophical case for how the very success of physics became, for a while, an impediment to further success.

Cosmology and the Progress of Physics

As we may recall, one objection against my thesis was that we apply truly fundamental physics (e.g., particle physics) to settle issues in space astronomy and physics, not the other way around. This objection has two corollaries. First, our truly important scientific views are not likely to be affected in significant ways by what we do in space. Second, as a consequence of space science, some of our views of the universe may well change, but they are of such remote or esoteric phenomena that the opportunity for practical consequences, even in the long run, is slim at best.

Let us take up first the matter of significance. There is a pecking order in the natural sciences, with physics clearly at the top. This is the case for three reasons. One is historical: physics led the way in the Scientific Revolution and presumably set the standards for subsequent science. A second reason is that physics deals with processes that are fundamental to the natural world; it deals with what all objects have in common. It is not surprising, then, that important changes in physics tend to be felt in many scientific places. The third reason is that the mathematical and experimental rigor of physics, coupled with extraordinary feats of the imagination, maintains the very high prestige of physics.

Even if such high prestige is well deserved, there is an unfortunate tendency to think that other sciences have much to learn from physics but little to teach it. And so we not only give priority to physics but even within physics we may downgrade what the received wisdom does not consider fundamental. For several decades, the most pressing problems were thought to reside in the very small, for the simple reason that the smallest components of matter are presumed to be the building blocks of the universe. These problems are the province of particle physics, and for some years the leading view to explain the nature of particles has been the Standard Model. This model contains twelve basic particles, fermions, which come in two classes: quarks and leptons. The theory of quarks goes by the strange name of "chromodynamics." The name is strange because it appears to refer to the dynamics of color, but the "color" in question is a property of quarks that cannot be seen (charm is another whimsical property of quarks, and the name "quark" itself is a nonsense word from James Joyce's *Finnegans Wake*). The word "lepton" means light (in mass), as the electron, the most famous lepton, is supposed to be. But now there are "heavy" leptons. To these creatively named particles, proponents of the Standard Model have added the Higgs particle, presumed to give mass to the other particles. Quarks are the building blocks of hadrons (e.g., protons and neutrons) which together with electrons form atoms. The model also accounts for three of the basic forces: electromagnetic, and the weak and strong nuclear forces. One particular property of gravity is how weak it is compared to the other forces. Taking the value of the strong force to be 1, the other values are as follows: electromagnetic, 10^{-2}; weak, 10^{-6}; gravity, 10^{-40}. The Standard Model posits the existence of four force-carrying particles named bosons, which include the photon and the gluon, which "glues" quarks to other quarks.

The fundamental physics contained in the Standard Model, however, does not account for gravity, although the discovery of the Higgs boson, once it is more fully worked out, may explain why some particles have mass. Some theorists hope the Higgs boson will explain why gravity is such a weak force. Perhaps a careful study of the cosmic microwave background, making use of the James Webb Space Telescope and others of its generation, will show evidence that heavy Higgs bosons decayed and only the lighter ones survived. That might account for why gravity is weak. At any rate, the connection between the Higgs boson and cosmology is acquiring fundamental importance.[19]

Since the quest to develop the standard model further has normally relied on building ever bigger particle colliders, which provide for higher and higher energies, it is not surprising that many physicists want to build now a truly gigantic accelerator, the Future Circular Collider, 100 kilometers long, almost four times longer than CERN's gigantic Large Hadron Collider (LHC). As you can imagine, the practical complications would be overwhelming. But particle colliders that long, and even far longer, can be built on the Moon, under the surface to protect them from temperature changes, and powered by sunlight.[20] The dream is to build one that would circle the whole Moon and would produce 14 quadrillion electron volts—about 1,000 times more energy than the LHC. But I suspect that physicists eager to advance their knowledge of particles might be willing to settle for, say, 200 times more energy than the LHC.

A rather popular way to bring gravity and quantum physics together, until recently, has been string theory, which supposes that particles are incredibly small loops or strings. It requires the existence of many dimensions in its attempt to give a consistent account of quantum gravity. Unfortunately, that account does not yet have any empirical evidence to support it, nor does it propose many experimental tests that would allow us to acquire that evidence. Even more fanciful views would have many universes existing side by side, so to speak, although such views boast of no more evidence to their credit than plain string theory.

The importance of quarks, and later of strings, was that they presumably explained the variety and properties of subatomic particles. More specifically, they pointed the way toward a unified account of the basic forces that act between particles, which physicists often call the "basic" forces of nature. Fundamental research was, then, research about those forces. Not that particle physics is the only kind of research a respectable physicist can undertake. But other physics shines by reflected glory, so to speak, and thus the more removed from the center of the discipline the less important it was thought to be.[21]

Today, however, since we must deal with questions about the nature, origin, and evolution of the universe, space astronomy is clearly fundamental science by even the narrowest and strictest of criteria. Much rides on the telescopes that our rockets may take into the heavens. It should be clear also that the more details we know about the universe, the more hints we benefit from in trying to devise new theories to explain its origin and its evolution. Science needs to rub against nature, for such friction polishes and sharpens the rough guesses that humans make about their universe.[22] And as we can see now, at the edge of the universe we find the end of a journey through space science that begins here and brings us back.

Consider the notion that investigating black holes and other strange objects presents extraordinary opportunities to challenge much of physics. It has been said that in black holes all of physics comes to an end. The reason is that in a runaway gravitational collapse, as presumably happens in a black hole, matter and energy disappear into a single geometrical point at the center of the black hole. This point, called a "singularity," obviously contains no space. And it contains no time either since, as we will see later, time slows down in the presence of a strong gravitational field. Where the field is practically infinite, time simply does not "happen." But the laws of physics make no sense outside of time and space. Thus, we seem to have a situation in which matter-energy is no longer subject to the laws of physics as we can presently conceive of them.[23]

Moreover, another serious complication arises. As the matter and energy in the black hole collapse toward the singularity, a moment comes when they are compressed into a volume so small that quantum effects may begin to dominate. This would mean, for example, that the account of the previous paragraph could not be right, since the Heisenberg uncertainty relation between position and momentum would rule out any deterministic prediction about the behavior of matter in such a small volume. Indeed, it could be thought that a similar problem may show up at the beginning of the universe. The result is that we seem to have a conflict between the two main theories of physics: general relativity and quantum physics. A future compromise, quantum gravity, has been the goal of many theoretical physicists, particularly string theories, but without any success so far.

The full-fledged development of space astronomy and space physics will enhance in a myriad of ways the dialectical relationships between different levels of studying the universe. As I have already mentioned, in the solar system we find clues about the rest of the universe, and our general cosmology affects our ideas of the solar system. In a similar fashion, terrestrial physics and astrophysics have been intertwined since the beginnings of the Copernican Revolution. Space physics promises to make that relationship even more intimate. And this is all we need to answer the queries about the long-term consequences of the changes in points of view that space may bring about.

The future contribution of space science to this new approach to the experimental study of gravitation will not be independent of the explosive growth expected of space astronomy. An accurate determination of the relevant masses is essential to the study of these peculiar gravitational phenomena. But to make such determination, we must also determine, for example, stellar distances and magnitudes very accurately. This will be the minimum benefit from the more powerful and more varied astronomy that will supplement from space the work of astronomers on the ground. And, of course, we wish to examine the instances of massive gravitational collapse. Even though a black hole does not permit radiation to escape—the curvature of its disturbance of space-time is so pronounced that we might say a black hole is a bottomless pit—it nevertheless makes itself known by the collapse of the matter around it. The gravitational force is so strong that great energy is released as matter sinks into the black hole. That energy can be detected with X-ray telescopes, and so the search for strong sources of X-rays becomes greatly significant.

Space science is thus becoming a significant factor in the ingenious attempt to turn a very theoretical subject into an experimental science, and in some cases into a branch of engineering. We can see the emergence of a familiar pattern: theory leads to new practical concerns, and those practical concerns in turn put us in a position to test theory—sometimes the same theory, sometimes theories in other fields. It is this rather complicated connection between theory and practical concern that eventually forces theory to work under circumstances different from those of its origin. Our views of the universe have permitted us to build the spaceships and the instruments to challenge many other views that we also hold. Ultimately the ensuing transformation will lead to a new round of challenge, and the dynamic character of science will be enhanced. This task will be aided in no small measure by the other great advantage of space science: by escaping from the confines of our own planet, it removes barriers and gives us a larger perspective.

To pass up the opportunity to enrich our cosmology so immensely would be far more than folly for the scientist who wishes to understand the universe. In the *Republic*, Plato describes a group of men who are chained facing the back wall of a cave. By the entrance to the cave there is a road, and beyond the road a fire that projects on the wall the shadows of the objects that pass in front of the cave. The men spend their time trying to determine what those objects are from the shapes they see before them. One day a man is set free and turns around. In reaching the outside world he is at first taken with fright, but soon he adapts to the sunlight and marvels not only at the objects whose shadows he had seen before, but at the many that had not even crossed his imagination, let alone his line of sight. And the question is, would this man go back voluntarily to his chains in the cave? Would he be satisfied with the guessing games based on what he now knows are mere shadows?

In a certain sense the atmosphere and the gravity well of our planet have been our cave and our chains. One of the great space pioneers, the Russian rocket theorist Konstantin Tsiolkovsky spoke of the chains of gravity and spent his life trying to break them. He and the other space pioneers have made it possible for us to come out and see the universe as if for the first time. Perhaps even then the universe will remain a complete mystery to us. But can we as cosmologists afford to reject the chance to work in the light and instead take back our place in front of the cave wall?

Transformation by Challenge

One of the ways in which exposure to unusual circumstances serves as a catalyst in the process of conceptual transformation is that it can help turn theoretical science into experimental science. To illustrate how space science plays that role, let us consider Einstein's general theory of relativity. Perhaps the most important insight of this theory is that the geometry of space and time (space-time) is affected by the amount and distribution of matter (or mass-energy) throughout it. This point is difficult to grasp, but it can be illustrated by means of the following analogy. On the surface of a regular coffee cup draw two points and trace the shortest line between them. The length of this line is going to be independent of the amount of coffee in the cup. Similarly, we would expect the properties of space and time to be independent of the "stuff" that makes up the universe. But consider now that the cup is made of some highly elastic material (say, the rubber used in toy balloons). In such a case not only the length of the line but the very shape of the path will be determined by the amount of coffee and by the way we squeeze the liquid about in the cup. Just as the geometry of the cup is affected by the liquid in it, Einstein might say, the geometry of space-time is affected in a similar way by the matter in it. Thus, the paths that light follows (geodesics) will vary from region to region; and time will slow down in the presence of strong gravitational fields (for it will be as if light had fallen into a deep hole—and so it will take longer to come out of it—which means that even the most efficient clock, using light signals, would measure time more slowly).

Figure 1.1. The Pleiades Star Cluster (M45). Located about 444 light years away, M45 is one of the closest star clusters to Earth. Although easily visible to the naked eye, Galileo was the first to observe the cluster through a telescope and found many more stars too dim to see without visual aid. The blue color is light reflected off the surrounding dim dust from the hot blue stars of the cluster. This image is about ten hours of data captured with a Celestron RASA 8. (Image courtesy of Michael R. Shapiro)

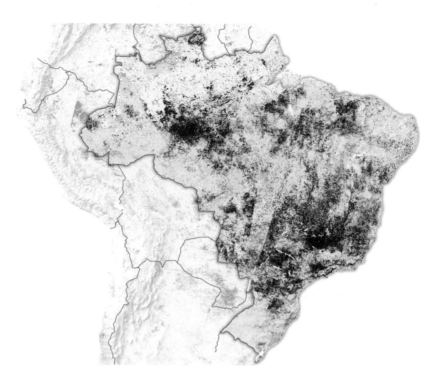

Figure 2.1. LANDSAT images help keep track of drought in Brazil. (Image courtesy of NASA)

Figure 2.2. A version of O'Neill's Island One, based on a Bernal Sphere. Rings on top of sphere would be individual torus areas probably devoted to agriculture. Larger versions could house millions of humans. (Image courtesy of NASA)

Figure 3.3. The Andromeda Galaxy (M31) is our sister galaxy in our local galactic group. Located approximately two million light years away, M31 is a spiral galaxy believed to house a supermassive black hole at its center. Although similar in size to the Milky Way, M31 is twice as massive, containing approximately one trillion stars. Many areas of nebulous star formation are visible in the image. This image is about twenty hours' worth of data shot with a Celestron RASA 8. (Image courtesy of Michael R. Shapiro)

Figure 4.1. The Earth-Moon system. (Courtesy of NASA)

Figure 4.3. Plume of erupting volcano in Io. (Courtesy of NASA)

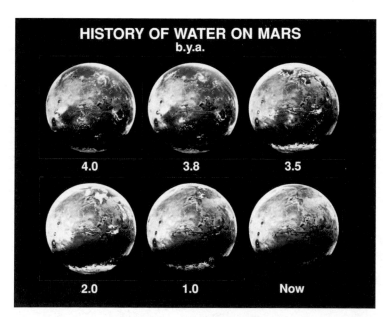

Figure 4.4. History of water on Mars. Numbers represent how many billions of years ago. (Courtesy of NASA/Michael Carroll)

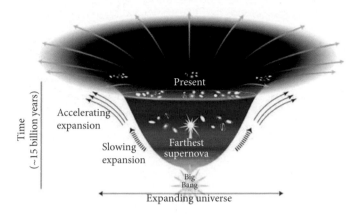

Figure 5.1. This diagram reveals changes in the rate of expansion since the universe's birth 15 billion years ago. The shallower the curve, the faster the rate of expansion. The curve changes noticeably about 7.5 billion years ago, when objects in the universe began flying apart as a faster rate. Astronomers theorize that the faster expansion rate is due to a mysterious, dark force that is pulling galaxies apart. (NASA/STSci/Ann Field)

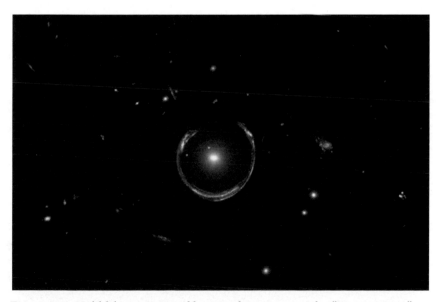

Figure 5.3. Hubble's gravitational lensing shows spectacular "Einstein Ring" around galaxy NGC1275. (ESA/Hubble & NASA)

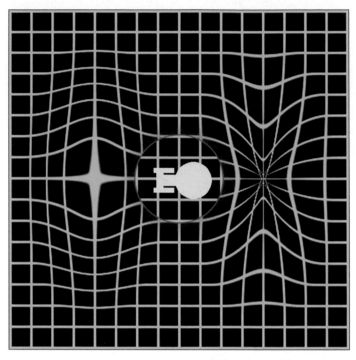

Figure 7.1. A starship has space-time expand behind it and contract in front of it. Eventually it may move faster than light, although it does not move relative to its local frame of reference. Its motion relative to other regions of the universe is the result of the properties of space-time. (Illustration by Trekky0623. Public domain image)

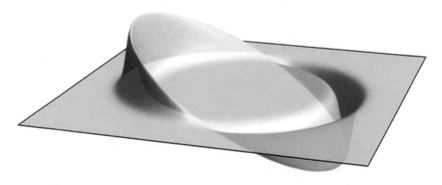

Figure 7.2. Regions of expanding and contracting space-time behind and in front of a central region. This combined action may move that central region (with a starship or a galaxy in its center) to speeds exceeding that of light. (Illustration by AllenMcC. Public domain image)

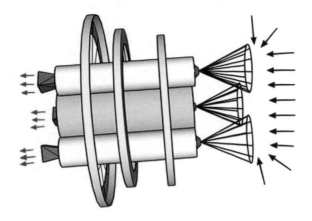

Figure 7.3. A space colony propelled by three Whitmire ramjets at relativistic speeds to begin human interstellar migration. Notice that the electromagnetic scoops in front of the ramjets create fields that take in all ions in the path of the ship. (Illustration by Ruoyu Huang)

Figure 10.1. Comet C/2020 F3 Neowise. The brightest comet since comet Hale-Bopp in 1997, Neowise was visible to the naked eye in the early evening near the Big Dipper. It was only discovered in March 2020 and reached its closest approach to Earth in late June of that year. This image is about one hour of data shot with the Celestron RASA 8. (Image courtesy of Michael R. Shapiro)

Figure 10.2. The Great Orion Nebula. Other than the Moon, this is the most photographed object in the night sky. The Great Orion Nebula is located about 1,400 light years away in the constellation Orion. In contrast to most nebulae, it is extremely bright, being visible to the naked eye as the middle "star" in the sword of Orion. The brightest area, known as the Trapezium, is home to intense star formation. The hot, young stars energize the surrounding medium and create the structure of the surround nebula. This image is about twenty minutes of data shot with the Celestron RASA 8. (Image courtesy of Michael R. Shapiro)

To put it in a nutshell, the theory claims that the geometry of space-time tells matter how to behave and that matter tells space-time how to curve. Massive objects thus distort the geometry of space-time (this is the effect of gravitational fields). And such geometry (or "shape," to state it colloquially) determines, for instance, how light can move through different regions of the universe. Near the Earth, space-time is almost flat, and thus it is very difficult to perform experiments to test Einstein's insights. As a result, for a long time there were only three tests of the theory. The most famous checked the prediction that stars directly behind the Sun should appear shifted to the side of the Sun. This should be expected, if the theory is correct, because the mass of the Sun carves a sort of a basin out of the fabric of space-time (Figure 5.2). Light coming very near the Sun would follow the contours of the basin, and thus be deflected; but it would also be "delayed," for it now has longer to travel. Unfortunately, the sunlight does not permit a check of this prediction except during eclipses. Even then tests were difficult. Nonetheless, they favored Einstein's theory, since during eclipses the stars in very close proximity to the Sun were seen to have shifted away from it within the range of Einstein's prediction.

With the advent of space exploration, however, far more precise tests become almost routine. Einstein's prediction is not really limited to light, but extends rather to all electromagnetic radiation, of which visible light is only one form. Thus, we can test Einstein's theory by tracking the radio signals of satellites that circle the Sun, or the radio signals from landers on the surface of Mars when Mars goes behind the Sun. That is, we can check not only the

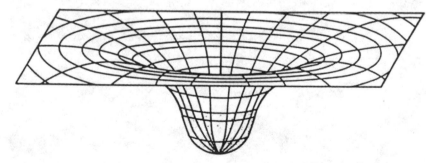

Figure 5.2. Strong force from a massive object (like a star) bends space-time. (From *A Theory of Wonder* [2021]. Reproduced courtesy of Vernon Press, Wilmington, DE)

prediction about the deflection of electromagnetic radiation but also the prediction about its delay.

The deflection of light by gravity sources is commonly called "gravitational lensing" (see Figure 5.3). When it occurs far beyond the amount of deflection that we could attribute to regular matter, then this much greater effect must be produced by invisible mass, that is, dark matter. Thus, gravitational lensing can give us a fair idea of the distribution of dark matter. Moreover, by looking further into time (the more redshifted light, the older the source), we can see how the distribution of dark matter has evolved and affected the evolution of the universe. If some of the models cosmologists have created of such evolution do not agree with the actual findings, then new discussions ensue about the nature of dark matter (for some astrophysicists, the motivation to develop their alternatives to dark matter). In any event, probably no theory in science has spun consequences so mind-boggling in number and in character as Einstein's general relativity.

Another such consequence is that gravity slows time (for clocks should then move faster the further away from massive bodies). Although it is possible to make this last kind of test using only terrestrial science, space offers much greater sophistication and permits far more ambitious developments

Figure 5.3. Hubble's gravitational lensing shows spectacular "Einstein Ring" around galaxy NGC1275 (see color plate). (ESA/Hubble & NASA)

along these lines. Already it permits us to use very precise clocks in spaceships whose distance from several points is determined to high accuracy by tracking stations using lasers. The advances in chronometry are respectable enough that the entire field goes beyond applied science into engineering.[24]

Of extraordinary importance also has been the discovery of gravity waves in 2016—one hundred years after Einstein predicted them,[25] for they provide an entire new window into the universe since matter is transparent to them. The feat was performed at the underground Advanced Laser Interferometer Gravitational-Wave Observatory (LIGO), which has had fifty more detections since then. LIGO works in conjunction with space telescopes that keep track of black holes, neutron stars, and potential locations of dark matter. The most sophisticated is the new German-built space X-ray telescope named eROSITA, launched in 2019, and capable of imaging the entire sky.

A great improvement on LIGO has been proposed in a new study: the Gravitational-Wave Lunar Observatory for Cosmology (GLOC). On a permanent Moon base, generations and generations of scientists could study sources of gravitational waves within 70% of the observable universe. GLOC would have two arms 40 kilometers long each, ten times longer than LIGO's, and would be extremely sensitive.[26]

Another proposal is for a space-based gravitational wave detector: the Laser Interferometer Space Antenna (LISA), which will use three spacecraft arranged in an equilateral triangle in orbit around the Earth. Each arm will be 2.5 *million* kilometers long.[27] LISA, to be launched in the 2030s, will have a Chinese companion, TAIJI, which presumably will be more sensitive to very high and very low frequencies.[28]

The Apollo astronauts set up a "crucial experiment" between general relativity and Dicke's theory of gravitation, which predicted that the Moon's orbit is deflected toward the Sun. The predicted change in that orbit was so small, however, that a test was out of the question until the landing of Apollo 11 on the Moon. The first humans on the Moon installed a reflector (the first of a series now in place) for a laser beam from Earth. This special equipment permitted extremely precise measurements of the distance between the Moon and the Earth, thus setting up an experiment that pit Dicke's theory against Einstein's. Einstein's won.[29]

I have given summary examples of many important experiments. To give a glimpse of the incredible precision and sophistication of space experiments, I will now be more detailed in my description of Gravity Probe B, the orbiting gyro experiment directed by Francis Everitt of Stanford University. This experiment tested general relativity in two most significant respects (it was originally scheduled to be launched by the Space Shuttle in the late 1980s, but it was not placed in orbit until 2004–2005). According to Einstein's view, the distribution of mass determines the geometry of space-time. But when a big mass rotates, it places space-time in motion as well; that is, it drags space-time along. To make a very precise calculation of the drag value, we have to know very precisely the values of the mass and the rotation involved. These values have been determined for the Earth, which also has a mass big enough to permit, according to the theory, a small but detectable amount of space-time drag. Four small, nearly perfect spheres—gyroscopes—placed in polar orbit had their rotational motions affected by such a drag. The object of the experiment was to measure that effect on the rotation—more specifically, the precession of the spheres with respect to the fixed stars—and see how it agreed with the values predicted by the theory.

Everitt's experiment also aimed to measure a most startling prediction of the general theory of relativity. When an electrically charged body rotates, it produces a magnetic field. Because of the structure of general relativity, we can ascribe a similar effect to rotating gravitational masses. That is, a field should be produced that would create torques on the orbiting gyroscope (this would lead to what is called the "gravito-magnetic precession of the gyroscope"). A positive result of this experiment would confirm the existence of magnetic-like properties of gravity, a discovery of much importance.

A great deal of advanced technology went into this experiment, involving sometimes a million-fold improvement in precision over previous technology. The four spheres were so perfect that if you blew one up to the size of the Earth, it would not show a deviation from the average radius larger than two feet. Such sphere had to be suspended without friction, and its minute precessions measured (one of 6.9 arc seconds per year, the other 0.05 arc seconds per year). These specifications can be met in space after considerable ingenuity and effort. The quartz spheres are coated with metal and suspended by the action of small electric fields. Their rotation was monitored thanks to the fact that since it was electrically charged and rotating it produced a small magnetic field.

On the Earth thousands of volts would be required to keep the spheres suspended, with many unwelcome side-effects added to a variety of disturbances typical of the environment—atmospheric, seismic, human, and so on. But even if all those things are accounted for, gravity alone would demand a degree of perfection in the spheres simply unattainable. The reason is that insofar as the spheres are less than absolutely perfect their individual center of mass will vary slightly from the center of a perfect sphere of the same radius. The stronger the gravitational field, the larger the rotational distortion produced by such a variation. Given the desired experimental performance—3×10^{-11} degrees per hour—under the gravitational acceleration of the Earth, the deviation in radius at any one point cannot be larger than 10^{-16} of the radius, which in this case it means that the deviation must be less than $1/1,000$ of a *nuclear diameter*. That is plainly out of the question. By contrast, in orbit the deviation had to be less than 10^{-6}, which was difficult but possible to achieve.

Everitt reported the following results: "Analysis of the data from all four gyroscopes results in a geodetic drift rate of -6601.8 ± 18.3 mas/yr and a frame-dragging drift rate of -37.2 ± 7.2 mas/yr, to be compared with the GR predictions of -6606.1 mas/yr and -39.2 mas/yr, respectively ("mas" is milli-arc second; 1 mas = 4.848×10^{-9} rad)."[30] See also Ronald J. Adler and Alexander S. Silbergleit, "A General Treatment of Orbiting Gyroscope Precession."[31]

The great precision and exhaustive character of our exploration of the cosmos will help give us also a better understanding of the evolution of planetary systems and their stars. It will, thus, help us also understand the nature and evolution of the solar system, which will result in a better understanding of the Earth and the Sun. In this respect, the new astrophysics and cosmology dovetail with the concerns discussed in Chapter 4.

Apart from gravitational lensing, discussed earlier, a most important investigation of the universe involves what is called the "Cosmic Web," which presumably gives us the distribution of dark matter. We can use these findings to account for how star clusters grow into big galaxies and galaxy clusters, as well as for the effects of gravity on collisions of galaxies, the speed of rotation of galaxies in the inner and outer parts of them, the rotation of galaxy clusters, and fluctuations of temperatures consistent with the cosmic microwave background. The consensus is that only dark matter can do the job of explaining these and other important features of the universe. A new study has provided evidence that the rotation of the central bar of the Milky Way has slowed down by as much as 24% since the formation of the galaxy, which explains why the stars of the Hercules stream have migrated over the past few billion years.[32] The usual measurement of dark matter determines its gravitational energy—indeed, that was the main reason to propose the existence of dark matter. But this new study adds a second important property: its inertial mass. It also adds more credence to the existence of a halo of dark matter around the galaxy. This finding is difficult for the alternatives to dark matter to match.

Gravitational waves may also help us determine the rate of the expansion of the universe, the first step in then trying to determine the nature of dark energy. There are two preferred methods for determining such expansion, with two different estimates. One estimates the distances of supernovae explosions and the pulsating stars called Cepheid variables, and then from those distances calculates the expansion rate. The other tries to determine the change over time of the cosmic microwave background to calculate how fast the universe has expanded. But recently many studies have carried out estimates of the expansion rate based on different approaches. Using gravitational waves from neutron star mergers and other phenomena is a very intriguing new candidate.[33] Of course, having an accurate determination of the expansion rate seems to be a requirement for attempts to then determine the nature of dark energy. When it comes to dark energy, as with dark matter, some may express doubts. In Chapter 7, I will suggest a modest proposal of my own.

Space physics and space astronomy have thus been shown not to be exceptions to the theses on the nature of science that I advanced in Chapter 3. Some of their activity completes the task that justified comparative planetology (e.g., solar physics and the astronomical efforts to put it in the proper context). And some other was shown, against the objection, to be fundamental science, and therefore it should be presumed to exhibit the natural connection with serendipity used to justify scientific exploration in general.

I cannot deny that much of what space science proposes to do sounds very esoteric. But so have sounded nearly all the revolutionary advances in the history of science. In some instances, the masses and energies that we wish to study with the aid of space are as large as the effects that we wish to measure and are small in other instances. How could they be of practical relevance? Equivalent questions to those we ask now about the relevance of general relativity, for example, could have been asked earlier of the special theory as well. Relativistic effects become pronounced only at extraordinary velocities (close to the velocity of light, 300,000 kilometers per second). But what regular person is ever going to travel at that velocity? Someday we might, actually, but the point is that those strange effects do show up in particle accelerators and make their way into our contemporary physics. Indeed, much of contemporary physics is based on the study of phenomena so small as to be beyond the conscious experience of any regular person. The grand man of atomic physics himself, Niels Bohr, often remarked that it was pointless even to ask questions about the reality of the processes of micro-physics. Nevertheless, the study of the extremely small and the extremely fast has produced surgical lasers and many other beneficial wonders in our time. We have no reason to expect fewer rewards as our playful science moves into the cosmos.

Notes

1. "News Focus: In the Afterglow of the Big Bang," *Science* 327 (January 1, 2010): 27.
2. researchnews.osu.edu/archive/astrotherapy.htm
3. ATLAS and CMS Collaborations, "Combined Measurement of the Higgs Boson Mass," *Physical Review Letters* 114 (2015): 33. doi:10.1103/PhysRevLett.114.191803.
4. *NASA Science*: Dark Matter, Dark Energy. https://science.nasa.gov/astrophysics/focus-areas/what-is-dark-energy
5. T. S. Kuhn, *The Copernican Revolution* (Harvard University Press, 1957).
6. Beginning with Classical Greek science and lasting into the development of chemistry in the modern era, natural philosophers believed the world was made up of four basic elements: earth, water, air, and fire. These elements were confined to sublunary space. Beyond the Moon, Aristotle thought, space was made of a transparent fifth element (quintessence) that had a natural circular motion (the perfect motion) unlike the natural vertical motion of the terrestrial elements.
7. S. Toulmin and J. Goodfield, *The Fabric of the Heavens* (1961), p. 143.
8. Toulmin and Goodfield, *The Fabric of the Heavens*.
9. Toulmin and Goodfield, *The Fabric of the Heavens*.
10. S. G. Brush, "Planetary Science: From Underground to Underdog," *Scientia* 113 (1979): 771–787.
11. Brush, "Planetary Science."
12. Brush, "Planetary Science."
13. Galileo Galilei, *Dialogue Concerning the Two Chief World Systems* (University of California Press, 1953), p. 328.
14. Galileo, *Dialogue Concerning the Two Chief World Systems*, p. 335.
15. Feyerabend.
16. Galileo, *Dialogue Concerning the Two Chief World Systems*, p. 335.
17. NASA Exoplanet Exploration. https://exoplanets.nasa.gov/
18. https://www.nasa.gov/directorates/spacetech/niac/2020_Phase_I_Phase_II/lunar_crater_radio_telescope/
19. M. Shaposhnikov, "The Higgs Boson and Cosmology," *Philosophical Transactions of the Royal Society A* 373 (2015): 20140038. http://dx.doi.org/10.1098/rsta.2014.0038

20. Emily Conover, "Physicists Dream Big with an Idea for a Particle Collider on the Moon," *Science News*, June 10, 2021.
21. For an account of the contemporary low status of the planetary sciences (until rather recently), see Stephen G. Brush, "Planetary Science: From Underground to Underdog," *Scientia* 113 (1978): 771.
22. This is a highly objectionable aspect of string theory: that it makes no contact with the universe we actually observe. Perhaps it will eventually, though.
23. A no longer popular suggestion was that of the "white hole," which would return to a different part of the universe the matter and energy that collapsed into the singularity of the black hole.
24. For an account, see Carroll O. Alley, "Proper Time Experiments in Gravitational Fields with Atomic Clocks, Aircraft, and Laser Light Pulses," a lecture published in *Quantum Optics, Experimental Gravitation, and Measurement Theory*, ed. P. Meystre and M. O. Scully (Plenum, 1982).
25. Davide Castelvecchi and Witze, "Einstein's Gravitational Waves Found at Last," *Nature News*, February 11, 2016. doi:10.1038/nature.2016.19361. S2CID 182916902
26. K. Jani and A. Loeb, "Gravitational-Wave Lunar Observatory for Cosmology," *Journal of Cosmology and Astroparticle Physics*, June 24, 2021.
27. See NASA: LISA: Laser Interferometer Space Antenna. https://lisa.nasa.gov
28. "China Unveils Plans for Two New Gravitational-Wave Missions," *Physics World*, July 11, 2018.
29. N. Calder, *Einstein's Universe* (Viking, 1979), pp. 97–98.
30. C. W. F. Everitt et al., "Gravity Probe B: Final Results of a Space Experiment to Test General Relativity," *Physical Review Letters* 106 (2011).
31. R. J. Adler and A. S. Silbergleit, "W. H. Hansen Experimental Physics Laboratory document," February 7, 2008.
32. Rimpei Chiba and Ralph Schönrich, "Tree-Ring Structure of Galactic Bar Resonance, *Monthly Notices of the Royal Astronomical Society* 505, no. 2 (August 2021): 2412–2426. https://doi.org/10.1093/mnras/stab1094
33. Stephen M. Feeney, Hiranya V. Peiris, Samaya M. Nissanke, and Daniel J. Mortlock, "Prospects for Measuring the Hubble Constant with Neutron-Star–Black-Hole Mergers," *Physical Review Letters* 126 (April 2021).

6

Space Biology

The well-being of human life has been the ultimate justification for the physical branches of space science. But can we justify the space science of life itself? What is the value of doing biology in space? Many observers, space scientists included, do not think highly of the scientific prospects of the field. The main purpose of this chapter is to meet their objections, to explain how the study of life in space also shows the sort of serendipitous, deep practicality shown by the other space sciences.[1]

Space biology can be roughly divided into two main areas. The first investigates the possibility of extraterrestrial life; it used to go by the name of exobiology, but it is now called *astrobiology*. The second investigates the behavior of *terrestrial* life in outer space; it corresponds to the idea most laypeople have of space biology.[2] Since these two areas are distinct, they will receive separate analysis and separate justification in each half of this chapter.

Astrobiology

It used to be said with derision that astrobiology is a science without subject matter. We have never found any alien life, and for all we know terrestrial life might be the only life in the universe. Or even if there is life somewhere else, we might never find it. Or if we stumble upon it by chance, we might not recognize it as life.

Astrobiology did have its moment of fame in 1975 with the Viking missions to Mars; but, after the Viking landers failed to find life, the field steadily declined in prestige and seemed moribund until two extraordinary developments resuscitated it.

The first was the dramatic advance in the search for planets around other stars. In the last few years thousands of extraterrestrial planets have been discovered, including many rocky planets.

The second extraordinary development was the discovery of organic compounds in a Martian meteorite (ALH84001), and the tantalizing

The Dimming of Starlight. Gonzalo Munévar, Oxford University Press. © Oxford University Press 2023.
DOI: 10.1093/oso/9780197689912.003.0006

suggestion that some worm-shaped structures found inside might be fossils of extremely small Martian bacteria. The monumental excitement created by the announcement led to a very ill-tempered controversy in the 1990s. Both sides would agree, however, that by comparing terrestrial and extraterrestrial life we would learn much about our own life and our own planet.

After discussing this issue, I will examine the negative results of the search so far and evaluate the hope created by the Martian meteorite. And, finally, I will determine why it makes sense to continue doing astrobiology even in the face of uncertain, or even negative, results.

The Motivation for Astrobiology

What would extraterrestrial life enable us to learn about our own kind of life? The answer is clear. All life on this planet is based on the same carbon chemistry and apparently all have the same genetic code. Of the many possible amino acids, only twenty are used to build proteins. DNA, the reproductive code for terrestrial life, makes use of only four bases (although it has been recently discovered that several bacterial viruses substitute in a fifth base).[3] Moreover, organic molecules can be left-handed or right-handed, but terrestrial life prefers left-handed amino acids and right-handed sugars. Are these circumstances mere accidents of organic evolution, or are there fundamental reasons why life has taken these particular turns on this planet? Even *one* other kind of life would permit us to make great strides in examining these matters. For that other life may use a wider range of amino acids and bases, or it may prefer right-handed amino acids or left-handed sugars. One result of such a finding may be that, say, a particular chemical balance in the Earth's early oceans caused the preference for left-handed amino acids. Or the alien life may be similar to life on Earth, which would reveal to us some sort of organic inevitability. The new perspective would be very fruitful in trying to understand our own biology at all levels.

It would be especially useful to observe stages of organic evolution and to study life as it begins in a new world, or at least to find fossil records of such beginnings. Beyond the stage of primitive cells, radically different alien forms of life would still offer great rewards, as we will see later. Even if perchance organic evolution produced in two similar planets similar primitive cells with essentially the same genetic code, the subsequent evolution would have much to teach us, for life in those two planets would undergo different histories of adaptation. Imagine, for example, that the now famous Alvarez asteroid had not crashed on the Earth. No one knows how dinosaurs would have continued to evolve, but it is possible that their grip on the surface of the planet would have been further strengthened. Mammals might have been thus forever condemned to crawl and scratch in the night like so many other vermin.

Even similar planets are likely to exhibit different tectonic histories. Plate tectonics brings continents together or breaks them apart; it throws chains of mountains up over the landscape; and it creates volcanoes where the plates rub against each other. In doing so, it brings some habitats to an end and others into existence. It destroys. It influences. It changes life in many ways. Consider how the variation in the size of landmasses influences the fauna and flora of a planet. Certain large animals, for example, need a large environment in which they can roam for long distances. Elephants used to travel many hundreds of miles in their annual migrations. As these big mammals went along, they ate a variety of plants, thus ensuring a balanced diet and permitting the vegetation at every feeding stop enough time to recuperate. Their considerable droppings were recycled, in the meantime, by armies of insects and bacteria. In a much smaller environment the vegetation would have been devastated, and the elephants would have suffered from poor diets and the unsanitary rot of their own excrement.[4] They would not have been fit.

Slight differences at the beginning of the history of a planet would alter the make-up of the crossroads that life has to face, first at the level of organic chemistry, and then at the level of cells—presuming that cells are common to living things. A eukaryotic cell (a cell with a nucleus) may well be the result of symbiosis between different varieties of prokaryotic cells (without a nucleus).[5] For example, the mitochondria in eukaryotic cells (see Figure 6.1) may be the remnants of prokaryotic cells that discovered how to use oxygen for energy and were swallowed but not digested by larger bacteria. Since eukaryotic cells are the building blocks of all complex organisms on Earth, we can imagine that different symbiotic relationships between primitive cells might have led to forms of life vastly different from those of our acquaintance. On planets so endowed, the subsequent interaction of life with the rest of the environment would have a multiplier effect, for they would change their environment in novel ways, and those new environments would lead life to adaptations that on Earth might meet only with misfortune.

Acquaintance with such alternative biotas would inevitably lead to profound transformations in biology, since biology would grow, and scientific

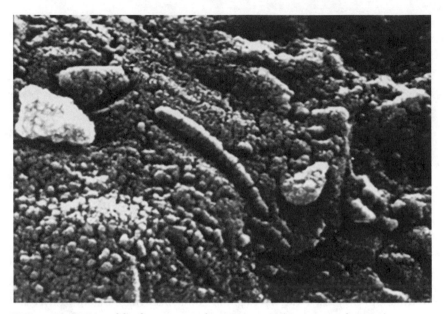

Figure 6.1. Worm-like features inside ALH84001. (Courtesy of NASA)

knowledge seldom grows without changes. In this, scientific knowledge resembles animals. Mammals, for example, did not just get bigger after the extinction of the dinosaurs. As their size increased, the structure of their skeletons had to change to accommodate their larger weight. In a planet with gravity similar to ours, a dog the size of an elephant would probably look much like an elephant. In an analogous manner, a science of biology that was suddenly much larger in subject matter would have to grow connections and supporting structures for which there was little need in the days of a single biota, similar to what we saw in the physical sciences in the previous chapter. And, of course, a radical new biology would have extraordinary implications for medical research and the new sorts of treatments that would result from that research.

The (So Far) Unsuccessful Search for Life in the Solar System

Given this apparent potential of astrobiology, then, it is not surprising that the Viking missions to Mars in the 1970s created great hopes of a scientific bonanza. In those missions, we could for the first time examine a world where life might exist. Moreover, the harsh Martian environment offered the possibility that if life had evolved there it would be radically different from Earth's. Thus, when the first experiments carried out by the robot landers seemed to indicate that photosynthesis was going on, which presumably only living things can perform, the excitement was deservedly extraordinary. But the excitement soon led to puzzlement when a chemical experiment determined that the soil possessed no organic carbon (C-H bonds). The fundamental chemistry of life simply was not there. And, therefore, neither was life. Now, this is the way the story was told at the time and for many years after. Today, several developments have changed the picture considerably.

The result of the Viking chemistry experiment was very disheartening at the time. We find organic matter in meteorites, in comets, even in deep interstellar space. How can Mars be so utterly deprived? Experiments by Stanley Miller, Harold Urey, and others have shown that some organic molecules form easily in the presence of gases that might have existed on both the Earth and Mars shortly after their formation (essentially, Miller mixed methane, ammonia, and hydrogen in a flask, subjected the mixture to electrical discharges, and then analyzed the goo that grew in the water, which yielded some amino acids and other organic compounds).[6] It is possible that in Mars those gases did not meet in any convenient ratios, but it must be said that the nature of the Earth's primordial atmosphere is a matter of dispute. In some scenarios, that atmosphere was dominated by CO_2, which would not respond to a Miller-Urey type of experiment by producing organic compounds. The terrestrial air of today, to mention another illustration, when subjected to discharges of electrical energy, produces smog, not the organic soup of the classical Miller-Urey experiment. Even so, if we drastically reduce the free oxygen and vastly increase the concentration of hydrogen in today's air, so as to have a reducing atmosphere containing also ammonia (NH_3) and carbon monoxide (CO), organic compounds may still form in a Miller-Urey type of experiment.[7] Why was life so unlucky on Mars?

At the time, many suspected that the Viking missions did not offer such a grand opportunity for astrobiology after all. The density of the Martian atmosphere is so low that liquid water could not now exist on it. This presents at least two problems for life. The most obvious one is that it is difficult for life to get by without liquid water—although perhaps the permafrost detected on the surface of Mars might suffice to sustain certain kinds of microbial life under the rocks.[8] The second is that when the permafrost is heated by sunlight it becomes water vapor. As the water vapor rises in the atmosphere, it is subjected to ultraviolet radiation, which disassociates the molecules of water. The freed light hydrogen atoms are eventually lost to space (Mars is surrounded by a gigantic envelope of hydrogen). The heavier oxygen could be expected to react with the substances in the soil. As we saw in Chapter 4, oxygen was a poison to earlier forms of life, precisely because it reacts so easily to form compounds. The oxidizing Martian soil would make it very difficult for organic evolution, let alone life. To make matters worse, Mars has no ozone layer and thus the ground is constantly bathed by ultraviolet radiation. All in all, some even claimed, the surface of Mars is far more antiseptic than the most fastidious operating room on Earth.

Furthermore, the Martian atmosphere shows no signs that its chemistry is being altered by the presence of living organisms. The Earth's atmosphere, by contrast, has far less carbon dioxide and far more oxygen that can be accounted for by physics and chemistry alone. Indeed, as we saw in Chapter 4, the composition of the Earth's atmosphere indicates that it is not in chemical equilibrium. And according to Margulis and Lovelock, the atmospheric gases can be kept so far from chemical equilibrium only through the action of living organisms, especially that of bacteria.[9] On the basis of this reasoning, by the mid-1960s Lovelock had predicted that no life would be found on Mars.

But is life on Mars truly impossible? Some astrobiologists argued at the time that the results of the Viking experiments do not warrant such a conclusion. They tried to meet the objections drawn from the chemical experiments in the following way. First, they pointed out that life would not alter the Martian atmosphere substantially if it only existed on the margins, so to speak. Small colonies of small organisms might sustain themselves without exerting pressures on the rest of the Martian environment large enough for detection with telescopes and the Viking state of the art. Second, the unprotected surface is not the place to look. There are sites in Antarctica devoid of life on the surface, but if we care to dig, we may find it in porous rocks below. Third, these astrobiologists brought up numerous examples of life surviving under extreme conditions: in the core of nuclear reactors, in underground streams with temperatures of hundreds of degrees Fahrenheit (or Celsius), under incredible pressures at the bottom of the ocean. Organisms have been found even deep in the Earth's crust. No extreme habitat, though, is as challenging as the Don Juan Pond in Antarctica, where the salinity is so great that a random sample is likely to fail the Viking test for the presence of carbon compounds. But living organisms exist there![10]

The claim of these astrobiologists was that the Viking experiments were designed to detect *average* life, whereas it is clear that if any life exists on Mars, it should be in *extreme* forms. Mars is an extreme habitat, if it is a habitat at all; experiments should be designed and interpreted accordingly.

If Mars fails us, life might have nevertheless made a stand in the organic clouds of Jupiter or perhaps in some underground caves in active Io. Another moon of Jupiter, Europa, is covered by smooth ice, and sometimes by significant amounts of water vapor, which indicates a good amount of internal heat, and probably an ocean of water under the ice. Smaller Enceladus, a moon of Saturn also covered by ice, has geysers that spew out not only water but even methane, which some scientists take as possible evidence of life.[11] Another notable prospect is Titan, the large moon of Saturn with a dense atmosphere and at least traces of organic compounds. Unfortunately, Titan is too cold for life as we know it—cold enough (−288 degrees Fahrenheit) that the argument about the ability of life to emerge in extreme habitats begins to wear thin.

At any rate, I should point out, the "extremophile" evidence can show only that once life begins, it can adapt to very hostile conditions. But it does not show that life could begin in such conditions. These are two very different things. Let me illustrate this remark by means of an analogy. During a woman's pregnancy, many substances can be lethal to the developing embryo (e.g., alcohol, tobacco, and hallucinogenic drugs). The chances for the new life are greatly hampered under those conditions. Once the baby is born, the situation begins to change. Eventually it may grow into an adult who smokes, drinks, and abuses drugs, none of which are conducive to a healthy life, but none of which need be immediately lethal either, as they could be to the embryo. Life might not be able to make a start on a planet that would otherwise be exactly like today's Earth, but it surely has no trouble flourishing in it now.

The encouraging prospects in Europa, however, do not depend on extremophile hypotheses. Europa is 10% water, with an ocean under the ice thanks to its internal heating, and with plenty of organic carbon because of its location in an area of the solar system rich in organics.[12] Now, it is not enough to have around liquid water, simple organic compounds, and a source of energy. Metals such as iron, zinc, copper, nickel, cobalt, magnesium, and manganese performed catalytic and other crucial roles in terrestrial organic evolution, as they do in the normal functioning of cells today. Without most of those metals, the prospects for life in any of the outer moons would be very dim.

Fortunately, Europa is likely to have them, for it is one of the few outer moons with a density (3.0), close to that of our Moon (3.3). Indeed, the rocky core of Europa under the water sphere would have a density even closer to the Moon's. It is a good bet, then, that Europa could offer a similar variety of metals.

Let us return out attention to Mars. It seems that in our sister planet the angels are with the astrobiologists. For if the Martian atmosphere was at one time much denser, life might have indeed begun; and then it might have survived, gone hail or high water. These hopes are sustained to a large extent by the tantalizing possibility that, once upon a time, Mars was far warmer and wetter, a possibility indicated by surface features that resemble river deltas and by dendritic channels also similar in appearance to river systems on our planet. These channels presumably constitute evidence of running water.[13] As we saw earlier, in a thin atmosphere, such as the present Martian atmosphere, water goes from solid to vapor without first becoming liquid. Therefore, this evidence of running water is in turn evidence that the atmosphere was much denser once upon a time.

There are alternative hypotheses on the dendritic channels, though. According to one of them, for instance, occasional but pronounced tilts in Mars's axis of rotation would expose one of the poles to the full action of the Sun. If that were the case, the melting polar cap would provide enough pressure to permit water to run off and presumably form those surface features—all without the benefit of a dense atmosphere.[14] Photographs by the Mars Reconnaissance Orbiter, with ten times better resolution than any taken before, indicate that lava and wind-driven dust have run through those presumed river channels and gullies far more recently than water, even though catastrophic floods might have carved them once upon a time.[15]

Nevertheless, the preponderance of new evidence indicates that the Martian atmosphere was far denser once upon a time and that liquid water ran on the Martian surface. The most striking findings are those of Opportunity, a Mars Exploration Rover. Opportunity found, for example, salt deposits in a region called Meridiani Planum. According to Mike Carr, the man who wrote the book on water on Mars, it is clear that a large body of water existed in that region. It is also clear that the "water had to pass through the ground to pick up the dissolved ions that ultimately were precipitated out as salts."[16] This means that the water in that region (a lake or a sea) could not have been mere runoff from melted polar ice.

If we assume that planets similar to the Earth have similar beginnings—in this case similar distributions of organic materials, atmospheric gases, and sources of energy—and if we keep in mind that our own earliest fossils are about 3.5–3.8 billion years old, it seems plausible to suppose that life made a start in Mars a long time ago. If that is so, the possibility exists that in some regions of Mars we may find fossils of organisms that thrived in days when the atmosphere was denser and warmer. NASA and the European Space Agency have landed Perseverance, a new six-wheeled Mars rover, to look for signs of fossil life in the Jezero Crater, which is presumed to have held a lake and a river delta in ancient times. It is hoped that some rocks there will contain biosignatures such as the layered mats formed, on Earth, by photo-synthetic cyanobacteria. Of course, finding living Martian organisms, most likely microorganisms, is a possibility, however remote. And in a decade or so we might be exploring Jupiter's moon Europa for the specific purpose of searching for extant life in the large ocean under the ice cover.

Although the Martian fossils that Perseverance might find would not be as exciting as living organisms, they still would be invaluable in that they would permit us to compare our form of life with an alien one. We may also be able to draw some interesting lessons from a failed interaction between life and a planetary environment.

And future missions may be guided by the consideration that liquid water is likely to exist in underground deposits. The permafrost is presumed to exist to a depth of hundreds of meters, which suggests that some liquid water may be found in proximity to magma deposits and other sources of thermal energy. And, somewhat controversially, ESA's Mars Express Orbiter's radar suggests that large lakes exist hundreds of meters under the surface.[17] Martian life forms, if any exist, might not be quite as extreme after all.

By an apparently similar reasoning, some would suppose that Venus might have some bacterial life deep below its surface, or at least a fossil record of its extinct biota. Unfortunately, the violent processes that melted the surface some 600 million years ago would have obliterated even the hardiest of bacteria, or their fossils.

In the case of Mars, support for the hypothesis of life has been claimed by David McKay's team's analysis of the Martian meteorite known as ALH84001 (see Figure 6.1).[18] According to this analysis, the meteorite contains globules of carbonate, polycycle aromatic hydrocarbons (PAHs), magnetite and iron sulfides, and some intriguing structures that some believe might be the fossils of ancient Martian bacteria. We know that the meteorite, which was found in Antarctica, came from Mars because the air trapped inside exhibits the same mix of rare gasses that the Martian atmosphere has. Great care was taken to rule out the possibility that the organic materials found could be the result of contamination (and indeed the proportion of those organic materials increases toward the center of the meteorite). Nevertheless, the reaction to McKay's results, particularly by those scientists considered experts on meteorites, was extremely hostile.[19]

The motivation for the hostility was in no small part the fear of ridicule, of having the study of meteorites branded as another "cold fusion," the big scientific embarrassment of the previous decade. The main argument against McKay's analysis was based on Occam's razor, a principle named after William of Occam, the medieval philosopher who insisted that we should accept the simplest explanation available.

As it was often repeated during this debate, extraordinary claims require extraordinary evidence (the so-called Sagan standard), but hydrocarbons, the magnetite, and the minute worm-like features could be explained by ordinary inorganic processes. There is, then, no need to conclude that Martian life caused the phenomena found in the meteorite.

Occam's razor, however, does not rule against McKay's analysis. In ALH84001, he found a collection of three things in an extremely confined space (a few nanometers across): (1) typical bacterial food (hydrocarbons), (2) structures that look like typical bacteria, and (3) typical excreta of bacteria (magnetite and iron sulfides). One simple hypothesis, life, accounts for all these phenomena and the fact that they are closely packed together: Martian bugs ate the hydrocarbons and left the droppings behind. The inorganic-origins hypothesis requires at least three separate mechanisms and has little to say about why they are together in such a small space.

Moreover, inorganic magnetite forms at a temperature about three times higher than that apparently experienced by the Martian meteorite. And the magnetite in the sample, unlike that produced by inorganic processes, is of an extremely pure form, which on Earth is normally produced only by bacteria. It is far from obvious then that the inorganic-origins hypothesis is simpler. Occam might have smiled on the life hypothesis instead!

In any event, ALH84001 left little doubt that Mars has had organic carbon, and that is a great find after the discouraging results from the Viking experiments of the 1970s, which ruled out life on the grounds that there seem to be no organic carbon on the surface of Mars they examined.

The Value of Astrobiology with or without Specimens

There was a more telling criticism of the notion that those worm-like structures in ALH84001 were fossils, however: many of the structures were much too small to be able to carry out many important organic functions that are typical of cells—some of them fifty or even one hundred times smaller than the smaller bacteria known at the time.

The ensuing controversy serendipitously spurred interest in the possibility that the Earth itself may contain bacteria that small. The interest increased when it was realized that the methods for looking for bacteria would not have detected such terrestrial "nano-bacteria" even if they existed. Lo and behold: biologists soon claimed to have found many such varieties of bacteria, even smaller than the presumed Martian bugs, right here on our own planet! Such discovery, however, seems to have been short-lived. Later investigations revealed that some candidates to the title of nanobacteria are nonliving mineral structures (e.g., calcium carbonate crystals do mimic bacteria in some respects and even reproduce).[20] Although not as exciting perhaps, this finding is nonetheless an interesting discovery in its own right. Moreover, it has some practical medical importance, since those nanostructures are apparently involved in the formation of kidney stones.

The adventurous search for very small forms of terrestrial life continued, however, and culminated in 2015, when Dr. Jill Banfield and her team of scientists from the University of California at Berkeley (LBNL) announced in *Nature Communications* their discovery of ultra-small bacteria (250 nanometers), about the size of the larger Martian worm-like structures (see Figure 6.2 and Figure 6.3).[21] They also *sequenced* the ultra-small bacteria genomes, which leaves no doubt about the bacterial nature of these structures. And we may hypothesize that primitive RNA bacteria without DNA could be even smaller: about two or three times smaller than ultra-small bacteria: about the size of many of the worm-like structures in ALH84001!

Scientific exploration challenges our ideas and frees the scientific imagination. The avenues of investigation that are so inspired can lead to important scientific discoveries, as we have seen in this example. Thus, the scientific investigation of potential instances of alien life, even if ultimately no life is found, are still likely to bring about discoveries in biology that may in turn have important implications for medicine.

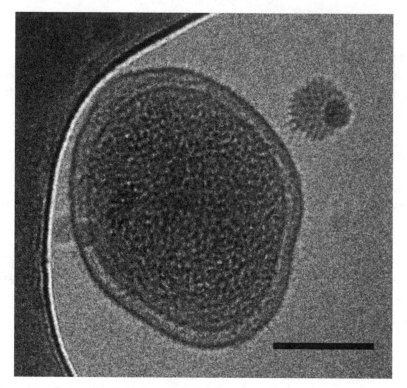

Figure 6.2. Ultra-small bacterium with dense interior and complex cell wall. Scale bar is 100 nanometers. (Drs. Birgit Luef and Jill Banfield. Reproduced courtesy of *Philosophy Study*)

Given such strong instrumental value, it is reasonable to conclude that our ethical obligation to preserve that alien life would be very high. It is an obligation to humanity, however, not to the alien life itself, at least not any more of an obligation that we have toward a group of terrestrial bacteria. Some may insist that the value of such life is intrinsic. It is worth pointing out, again, that the case made so far is purely instrumental. Moreover, the obligation to preserve that alien life would remain as long as it does not pose a threat to humanity. In that case, if we could not avoid it or contain it, we may have to destroy it. As Mark Twain put it, in *Letters from the Earth*, Noah could have rid the Earth of mosquitoes by keeping them off the Ark. In case of a deadly virus, for example, we would have to do far better than Noah.

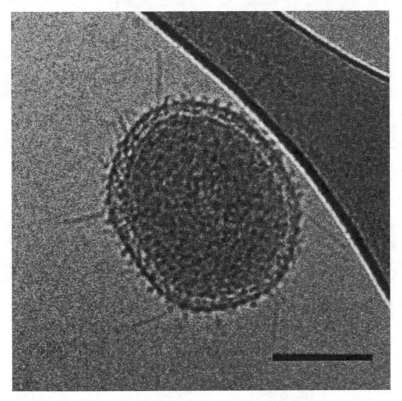

Figure 6.3. Hair-like appendages might be used by ultra-small bacterium to make contact with other microbes. Scale bar is 100 nanometers. (Drs. Birgit Luef and Jill Banfield. Reproduced courtesy of *Philosophy Study*)

Relationship between Astrobiology and Planetary Science

The lack of extraterrestrial specimens is an objection to the pursuit of astrobiology only if we accept a narrow definition of the field. Astrobiology goes beyond the search for extraterrestrial life: it is largely the application of space science and technology to understand how life may originate and evolve anywhere in the cosmos. As a matter of special interest, astrobiology often devotes itself to investigating how life originated and evolved on *this* planet. Astrobiology tries to determine, for example, what the Earth was like 3.5–4.5 billion years ago—what was the ultraviolet flux? What were the volcanic and other tectonic activity? How much molecular oxygen was in the atmosphere? And how much ozone? How much carbon, hydrogen, and nitrogen were "recycled" through the Earth's crust and how much was brought to the Earth by asteroids and comets? To decide these issues, we must go away from the Earth to study the older surfaces of the Moon and Mars, the presumably still primordial atmosphere of Titan, and the largely untouched chemistry of comets.[22] Astrobiology is thus inseparable from comparative planetology.

This connection is all the more evident when we remember that to understand the nature of an environment well we need to understand its origin and evolution. In the global environment of the Earth, life has played a crucial part (cf. Chapter 4). How life originated is thus a question of great importance if we are to understand how our global environment came to be as it is. At the same time, we cannot begin to settle that question without making some critical determinations about how the planet was formed, how its atmosphere was created, how much energy it received from the Sun in its early evolution, and in general all those questions that form an integral part of comparative planetology.

In trying to answer the question of the origin of life, however, there are great difficulties of substance and of method. For example, some investigators require explanations that make life somehow inevitable or at least very likely. Given the early conditions in the planet (e.g., a reducing atmosphere and later a primordial soup of organic materials), and processes that should be expected (e.g., radiation, lightning), organic evolution toward life should be highly probable. Another school of thought would have a series of extraordinary coincidences bring life about. Thus, even if organic matter was abundant on the early Earth, it would have taken an accident, or accidents, to get organic evolution on the road to life. To illustrate the sort of disputes involved, let me consider the hypothesis that a large Moon was needed for life to begin. Of course, if that hypothesis is right, we should conclude that life is probably very rare in the galaxy and not to be found in the solar system at all, except for our own kind.

According to this hypothesis, clays in shallow waters served as the templates for amino acids to combine into the first complex organic molecules by forming peptide bonds—bonds that can link carbon and nitrogen in separate organic chains. Such a bond can form when, for example, a nitrogen atom loses its bond with a hydrogen atom (H) and a carbon nearby loses its hydroxide bond (OH). Since the H and the OH combine to form water (H_2O), the formation of the peptide bond requires a loss of water. The problem is how to lose water in the presence of all that water in which the amino acids are suspended.

The tides created by the Moon provide the solution to the problem. When the tides go out, a residue of amino acids is left on the clays, and the heavier concentration in drier surroundings allows the peptide bond to form. Once formed, the peptide bond is stable in water. After many repetitions of this process, organic molecules of an increasing complexity can be formed. The function of the clays is to provide a mechanism for replication, but that matter will wait a few paragraphs. For the time being, the issue is the role of the Moon in the origin of life.

The Sun produces tides, too, but they are much smaller and perhaps not sufficient to bring about the needed peptide bonds. If a planet were closer to the star, it would enjoy greater solar tides, but unfortunately it may also become locked into a very slow rotation (the Moon always offers the same face to the Earth, Mercury rotates every 58.6 Earth days, and Venus's day is longer than its year). Having a day-night cycle seems important because the opportunity to move away from equilibrium gives the prebiological molecules a chance to vary, and this opportunity for variation is an essential characteristic of evolution. Cyclical events are in general favorable because they permit the molecules to reach a state of equilibrium to consolidate their gains before having to change again. Apart from the tides and the day-night cycle, we have the concomitant temperature fluctuations. And we have seasons because the Earth's axis is tilted in just the right way, thanks, again, to the Moon. Then we must also take into account the role of the magnetosphere and many other factors whose possible relevance or even their very existence may escape the experts at this time.

Insofar as any of these are large factors in allowing life to gain a foothold on Earth, the origin of life becomes an improbable event. But we simply do not know. As plausible as hypotheses such as this may seem today, they may sound very quaint in two or three decades, let alone in a few centuries. And even if the events in question were indeed factors in bringing life into our world, on further examination they may turn out to be just some among the many alternative mechanisms that could have provided for the evolution of ever-more complex molecules in a variety of other worlds. It happens all too often that when a mechanism cannot be immediately proposed to explain a particular step on the way to life, people who ought to know better jump to the conclusion that life on Earth was an extraordinary coincidence.

Many biochemists, for example, have felt that the problem of the origin of macromolecules is insoluble. At the most basic level, life consists of nucleic acids (such as DNA and RNA) that contain the genetic information, and of functional proteins (such as enzymes). If we imagine that DNA or RNA was the original macromolecule, we have to explain how it could replicate in the absence of enzymes, which are essential in modern living systems. On the other hand, if we imagine that the proteins came first, how could they have built around themselves the nucleic acids that would carry the information necessary for future coding of the same proteins?

For years, none of the mechanisms proposed seemed satisfactory, not even co-evolutionary mechanisms because the chemical association of nucleotides (the building blocks of nucleic acids) and amino acids (the building blocks of proteins) was just too problematic. On the face of this situation, some people thought that life was a stroke of luck. And some others even suggested that life had probably come from elsewhere to the Earth (panspermia), as if removing the problem of the origin of life a few light years amounted to a solution.

In 1973, Francis Crick and Leslie Orgel considered the possibility that life on Earth was the result of directed panspermia. From arguments given earlier, we can see that a policy of spreading life in all directions was likely to affect negatively the life already existing in planets around the galaxy. One factor that may have influenced scientists to advance fantastic stories decades ago was the concern for whether enough time had elapsed on Earth for the origin and evolution of life.

We have strong evidence of bacteria in rocks about 3.5 billion years old, and some evidence of fossils as old as 3.8 or perhaps 3.85 billion years old.[23] It is plausible that the Late Bombardment of comets and asteroids (around 3.9 billion years ago) may have sterilized the surface of the Earth. If life before then thrived in deep ocean vents and under the young continents, it might have survived and then spread anew once conditions were favorable again.[24] Otherwise, if life had established itself only on or near the surface of the Earth, it would have been annihilated. In the first case, organic evolution could have taken place over many hundreds of millions of years. In the second case, organic evolution would have had available less than a hundred million years to produce the miracle of life.

Let us consider the worst-case scenario. It is possible that organic evolution, which would lead to perhaps some primitive cells, resembles biological evolution in some key respects. In biological evolution we have an example of a process that may increase complexity while reducing the space of possible outcomes. Since the first characteristic is obvious, let me concentrate on the second. In general, evolutionary processes that have mechanisms for variation and selection can quickly reduce the space of possibilities. In his book *Why We Feel*, Victor S. Johnston describes a face-recognition program that he invented to "evolve" a criminal's face (police composites are not very satisfactory). In writing his genetic algorithms, he created dimensions such as width of chin, with several values, from very pointy to double chin, size of ears, amount of hair, and so on. Each "genotype" was made up of fourteen computer "genes" specified, all together, in a sixty-bit binary string. This "program can generate more than a billion billion different faces (2^{60})."[25] If a witness viewed one face every second, it would take him longer than thirty-six billion years to go through all of them. But assisted by Johnston's FacePrints program, a witness can come up with a close resemblance (far superior to a composite) to the criminal *in less than one hour!*

The program creates thirty random faces, and the witness ranks them (0 to 9) on how closely they "fit" the criminal's face. The program then creates a new generation of faces in which the "fittest" are represented in greater numbers (in proportion to their rank). This and future generations are produced by taking two selected genotypes at a time and mixing them, using crossover and a small number of mutations (in analogy to biological evolution). Moreover, David N. Reznick, an experimental evolutionary biologist, has done work on fish that suggest that the rate of natural selection may be forty to seventy times faster than that suggested by the fossil record (which, for obvious reasons, emphasizes morphology).[26] And very importantly, Lazcano and Miller estimated that the evolution from a primitive heterotroph with a small genome into a far more sophisticated cyanobacterium would take no more than seven million years.[27]

Organic evolution, if it is to parallel biological evolution, should then have mechanisms for variation and selection, apart from the specific requirements specified earlier. In a rich environment with a great many sources of energy, with many mechanisms for concentrating organic molecules, and so on that the early Earth would have offered, it would not at all be surprising that some characteristics would favor some molecules more than others, particularly in the presence of certain other molecules. Indeed, enzymes can make certain reactions millions of times faster than they would otherwise be. And as Christian de Duve argues, the chemical reactions of life have to happen quickly or the reagents may break down or dissipate, as they would in deep space, where the density of molecular clouds is low and the time scales are long.[28]

Let us consider now some plausible mechanisms that make the terrestrial origin of life look more likely than proposals for life arriving inside a meteorite and then spreading throughout. One of them, suggested by A. G. Cairns-Smith, is that microscopic crystals in clays can serve to replicate molecules. Such clays have a large capacity for adsorption, which causes tiny bits of proteins to stick to them, just as particles of meat do to the surface of a frying pan. The crystals in these clays would then grow and reproduce the patterns of the amino acids adsorbed in the clays.[29]

These processes can be repeated millions of times, until with the development of enzymes, as J. D. Bernal notes, we would also see the appearance of co-enzymes, some of which are identical to the nucleotides of RNA. As the co-enzymes are adsorbed, their efficiency in chemical energy transfer would give clear reproductive advantages to their associated enzymes. Under these conditions, a co-evolution of functional proteins and nucleic acids becomes possible. And this result presumably paves the road to the eventual origin of the first cells. This view has been buttressed by the work of the space scientist James Lawless, who has shown that clays do select precisely the amino acids that can form biologically active proteins.

There are many other hypotheses, some of them buttressed by experimental work, that fill in some of the steps deemed necessary to take us from atmospheric gases to living cells (e.g., proteins that can "make" their own RNA). But what is necessary and what is not depend on the approach one takes to explain the origin of life. First, there are different starting points. Some want, even demand, a reducing atmosphere (poor in oxygen, rich in hydrogen and other gases like methane). Others think that the original atmosphere was composed largely of carbon dioxide. Second, then comes a story of the evolution of organic matter, a story that may involve thousands of steps, of which only some are specified. And of course, there could be alternative plausible stories. What makes them plausible is that some of the steps that may have seemed baffling at one time can be produced in the laboratory now, while others can be explained theoretically. For example, the first generally accepted story gained its plausibility from the Miller-Urey experiment, in which a reducing atmosphere in a flask was subjected to electrical discharges. Presumably the result of the experiment was a soup containing the building blocks of life. Apparently, however, only very few interesting organic molecules were actually produced in such an experiment, and those were of very little complexity. The steps from there to, say, a self-replicating molecule, let alone a cell, are truly gigantic.

Since there are so many ways to tell the story, most equally unconstrained by the scant evidence, it is not surprising that the intuitions of different investigators differ on what is crucial and what is not. And even if most of the apparently necessary steps of a particular story can be accounted for by experiments, there remains the difficulty that the answer to one part of the puzzle is often at odds with the proposed answer to the next part (e.g., a molecule used as a building block for a more complex molecule is produced in an alkaline solution, but the more complex molecule has to be produced in its opposite, an acidic environment; this is not a fatal setback, since in living things the product of a reaction can be transported to a different internal environment to be used to build something else, and in general we find that natural processes have co-evolved in the living world to accomplish just this transport; cf. the example about elephants and their environments given earlier). The problem is that it all seems just too convenient. What we want to know now is not just how it could have been, but how it was—we want the "real" story.

To go from just-so stories to compelling hypotheses, we need a better understanding of the initial conditions on the Earth, and as we develop our hypotheses accordingly, we will get ideas of what sorts of evidence about the subsequent evolution of the global environment we may want to look for. One helpful way to proceed is to examine those worlds where, according to some approaches, life might have started, or at least where we should expect some small amount of organic evolution. To the extent that organic evolution has taken place there, we learn much about our own, once we factor in the relevant differences. To the extent to which organic evolution has not taken place, we also learn much about the failure of some forms of reasoning about the origin of life and perhaps get some clues about more appropriate forms of reasoning.

Astrobiology and comparative planetology will merge in many other contexts. Take, for example, the search for the origin of the organic carbon on the Earth, surely a needed background to make a definitive determination of how life started on the Earth. To have organic compounds, we first need to trace the carbon and the other relevant elements (hydrogen is normally easy, since it is almost everywhere, with exceptions such as the Moon and Mercury). We begin the search for carbon in the solar system and then see how it was apportioned to the Earth. If the Earth had a disproportionate amount of carbon, we must deal with a certain set of scenarios in which the Earth comes to occupy a privileged position. If carbon is very common in the solar system, as indeed it is, our scenarios are of a different sort, but we still want to know how the Earth came by the amounts that it has: did it happen during the initial accretion of the planet, or was most of the carbon brought in by the subsequent bombardment by comets and asteroids?

How do we trace the carbon in the solar system? We will look for clues in the comets and asteroids, as well as in the inner planets and all the other rocky objects of respectable size in the solar system, particularly the large moons of Jupiter and Saturn. If we wish, space exploration will make them all available to us. In the meantime, we may also examine the meteorites that we do have. In some of these meteorites we find that carbon is associated with two sets of rare gases—apparently the carbon serves as a casing that keeps the rare gases trapped. When the carbon is released from the meteorite, it comes out in two installments. In the first it also frees amounts of argon, krypton, and xenon in relative ratios that are pretty much the same as those found in terrestrial planets. Other carbon in some meteorites, however, encases a completely different mix of rare gases (of neon, xenon, and krypton) that cannot be accounted for by any process known in the solar system. Further investigation reveals that such is the mixture that can be expected in the process by which stars become red giants. This presumably leads to the conclusion that the stellar cloud from which the solar system formed was already seeded with carbon from the death of another star.

This interplay between biology and astrophysics takes place on many fronts. As another example, we may consider the matter of extinction, which has been described earlier. The twenty-six-million-year period for large extinctions was calculated by David Raup from the fossil records of marine animals. It was this periodicity together with the asteroid hypothesis of Luis Alvarez that led to the further astrophysical hypothesis that the Sun had a companion star (*Nemesis*, never found). Incidentally, the computer program used to estimate the effects of the Alvarez asteroid was based on a program originally developed by Brian Toon and others to study the dust clouds of Mars and, in turn, served as the basis for the so-called nuclear winter study— which some observers still consider a reliable model of the devastating nightmare the whole Earth would suffer in case of nuclear war.[30]

If such cycles of extinction are indeed determined by astrophysical events, posterity will have much reason to thank the day these combined biological and astrophysical studies were undertaken, even if the astrophysical causes turn out to be very different from those now proposed. But quite apart from preventing great disasters, the study of extinction cycles will contribute greatly to our understanding of the forces that affect the environment of the Earth.

These and other investigations underscore the intimate connections between astrobiology and those aspects of space science that deal with the formation and evolution of planets. Since the role of life has been of crucial importance for the Earth, and since we need to know specifically how terrestrial life may not only survive but also prosper, this study of origins is highly justified. To put the point differently, the biota and many of the other elements of the global environment co-evolved. Thus, to understand the evolution of one of those elements, we need to understand it in its relationship to the evolution of the others. Moreover, the very role of the imagination in trying to determine the range within which life can be born and the possible forms life may take provides a fruitful context in which to discuss questions of origin and evolution. For by the consideration of likely scenarios for life, and by the comparative examination of the planets in our solar system and of other planetary systems, we will be better able to understand not only how life came about but also why it took the paths that it did when it apparently had others available.

This serendipitous result of astrobiology, as valuable as it has been in giving us a new understanding of life on Earth, may pale in comparison with the creation in the laboratory of life forms that incorporate a twenty-first amino acid and others with nonstandard DNA codes![31]

It is clear that much needs to be done in this field, and that space science is particularly well poised to nourish its advance. At the same time, we should beware of placing unfair demands upon the field, particularly where it concerns the search for the origins of life. We should beware especially of the carefree use of probabilities in trying to settle this important issue. The most notorious is the estimate of the probability that all the constituent atoms of a cell may come together to form the cell all at once. Even for a strand of DNA the probability would be extremely low. And so it would be for any complex arrangement of matter, as long as we assume that it started from scratch. As Fred Hoyle put it, what is the probability that a Boing 747 will arise spontaneously from a tornado-swept junkyard?[32] Of course, the probability is close to nil. But this is a silly demand: cells are not formed from scratch. Some elements combine together more easily than others, and if they are abundant, then we will find many of their compounds. Such is the case with carbon and hydrogen. Once those compounds are formed, more complex compounds can more easily form (higher probability) using them, and so on. The rising complexity of molecules can give rise to very complex molecules indeed— and then the very long process of organic evolution is more likely to begin.

Robert Shapiro, a critic of the field, tries to impose two requirements that deserve special comment. He claims that the thesis that the origin of life was an accident is not scientific. Apparently, he feels that a truly scientific approach would explain why life was inevitable, given the Earth's early environment. And he also objects to laboratory simulations of hypothetical early terrestrial environments in which the experimenters manipulate the environment to determine whether certain complex molecules can be produced from certain others. He wants the experiments left alone, to see whether the molecule so produced is capable of evolving on its own (otherwise we are not really dealing with organic evolution, I presume). His suggestion is that much of the work in the field fails to meet these two requirements and he concludes that the field is in disarray.[33]

The first thesis is rather strange coming from a biologist. If organic evolution is evolution at all, it is subject to the vagaries of natural history. The evolution of mammals, for example, may have well depended on extraordinary accidents (such as the Alvarez asteroid, which made available to our ancestors the niches previously ruled by dinosaurs).

It seems that Shapiro is unhappy because the search for the origin of life does not demand the sense of inevitability that we expect from physics. But that is one of the differences between physics and historical sciences like geology or biology. Moreover, even if we wish to use physics as a model, it seems that either chaos theory or Prigogine's dissipative structures would serve us better. A very small change in initial conditions may lead to radically different outcomes. A tiny amount of a catalyst can produce an oscillating reaction (say where the color of the solution keeps changing from red to blue). At the time of the Cambrian Revolution (about 530–550 million years ago) there was a great explosion in the forms of life that began to populate the planet. An observer could not have predicted then that human beings were sure to come along millions of years hence, unless he had knowledge of all the accidents that would take place in the ensuing years, and of all the ways in which complex environmental relations were going to change. Nevertheless, this particular outcome of evolution (humans) is an accident, and so is any other particular outcome. If life is the outcome of organic *evolution*, life itself could be said to be an accident, too.

A compromise position may be defended. We need claim neither that life (as we know it) is an accident, nor that such life was inevitable. For example, we may *hope* for an explanation of origins that makes it look as if some accident of this sort (life) was likely to happen (e.g., a self-reproducing molecule that can protect itself from most typical, short-range, environmental dangers, even though its genetic code is very different from ours).

As for Shapiro's second requirement, it seems to me that we should want to create in the laboratory a molecule that can reproduce in the sorts of environments that we think may have existed long ago. It would be unreasonable to demand that such a molecule should reproduce in any environment that might develop if we leave the apparatus unattended. We must remember that most species that ever lived are now extinct. As the environment changed, only those organisms to which the change was not unfavorable were able to leave progeny. Thus, by a similar reasoning, a molecule may be of the right sort and still fail to reproduce under the conditions required by Shapiro instead of conditions similar to those that an evolving Earth made available to its complex organic molecules.

In its own ways, astrobiology thus illustrates how space science preserves the dynamic character of science in general. Its vigorous pursuit would inevitably lead to the profound transformation of our views of the living world. And since those views are linked to our understanding of the global environment, the resulting theoretical adjustment would be of great magnitude— and so eventually would be the change in the way we may interact with the universe. The justification of astrobiology is, then, ultimately much like that of the other space sciences, and in line with the general philosophical position of this book, whether or not we ever find a single extraterrestrial specimen!

The Prospects for Astrobiology

If we never find life, or fossils of life, in Mars, Europa, of anywhere else in the solar system, astrobiology will have to look at other solar systems for specimens. At first that seems to present very serious problems. A human mission to the stars would last thousands of years with the technology we can muster today. With the fancier technologies to be described in the next chapter we can get there much faster, but there is no serious commitment to them for the near future. There seems to be a shortcut available to astrobiology, and that is to make contact with intelligent extraterrestrial life. But that is a long shot, as we will see in Chapter 8.[34] Fortunately, space telescopes are coming to the rescue once again. As we have seen previously, life has a great influence in the composition of a planet's atmosphere, increasing the percentage of oxygen and methane, for example. And our new space telescopes will soon make almost routine the spectral analysis of the chemical composition of the atmospheres of exoplanets. Any indication of the presence of life in another star system will create great motivation to begin experimenting with technologies that might achieve the relativistic speeds that will allow us to take much closer looks in dozens as opposed to thousands of years. And let us remember that, given the number of exoplanets already found, is not too far-fetched to project the existence of billions of them in our galaxy, a fair percentage of which might be in the habitable zone, where planets may have liquid water on the surface.

How will we recognize life that might be different from ours? Do we have a scientifically acceptable definition of life? New tools have been proposed to identify life even if it is based on a very different biochemistry. The most prominent such tool is based on assembly theory, which distinguishes living from nonliving systems by their production in abundance of complex molecules that cannot occur randomly. This theory emphasizes what life does (produces very complex molecules) instead of attempting to define it. Presumably exobiology will behave similarly. The complexity of the molecules is quantified by an algorithm based on the number of bonds required to "assemble" the molecules. This "molecular assembly" (MA) number will be larger in biogenic molecules, for they require a very large number of bonds. After much testing, scientists are confident that only life can bring about molecules with very high MA numbers.[35] They have even determined a threshold to draw the line about life.[36] Incidentally, coming to understand the way in which living systems self-organize and assemble complex molecules may give us important clues about how to invent new drugs and other materials.

We have seen how the search for life in Mars and other places could have profound implications for biology and medicine. As a result, we have an ethical obligation to carry out the exploration of Mars and those other places in a manner that minimally disturbs them, so as to preserve such scientific treasures. Any possible large-scale colonization should wait for that scientific exploration to be completed, while greatly benefiting from its findings. The National Academy of Sciences, at NASA's request, has provided new guidelines for the biological burden to be placed on missions to Mars.[37] The emphasis is on reducing the risk of contamination by terrestrial life of the most likely environments for evidence of present or past Martian organisms. They include caves where ice water might be present, as well as aquifers very deep underground. As a result, the activities of missions to Mars should refrain from subsurface activities if ice has been detected within 1 meter of the surface. The landing site must be at a conservative distance from any subsurface access point to potential sites of astrobiological interest. Furthermore, every mission should be examined for how it might produce harmful contamination of Mars. These bio-burden requirements would make it unlikely that some group might be allowed to attempt to set up a colony on Mars, say, before the space scientific community is satisfied that our exploration of Mars has been reasonably thorough. These approaches are in keeping with the concerns about the priority of science expressed by James S. J. Schwartz in his book *The Value of Science in Space Exploration*.[38]

Terrestrial Life in Space

If a man hangs upside down for many hours, blood will rush to his head, his breathing will be impaired, and he will die. To prevent such a fate, evolution has provided us with means for telling which way is up. In mammals, these means include the otoliths of the middle ear, sophisticated muscular-skeletal sensing devices, and the coupling of eyesight with all these organs in the brain. This detection of gravity is no less important to a plant, which needs to send its roots into the ground in search of nutrients, and its shoots into the air in search of gases and sunlight. Since in orbit gravity is practically absent, many experts predicted that the perceptual and physiological disorientation would lead to heart trouble, depression, and mental impairment. The severity of these and other disorders would surely make crewed space flight impossible.

The march of events has confounded all these dire predictions about the fate of humans and other forms of life in space. Nevertheless, space flight does affect living things in a variety of ways. Caution has been called for, and caution has been exercised. The result has been a large body of research—mostly clinical research—aimed at ensuring the safety of astronauts and at establishing the degree to which humans can adapt to the low gravity and high radiation of space. The prominence of this body of research, however, has led to a distortion of the significance of doing biology in space.

Some observers have concluded, for example, that biological research in space makes sense only if we are planning to continue crewed space exploration.[39] Of course, they say, we have to know what space does to a human body if our astronauts are going to spend long times in that environment. And if plants and animals are to be an integral part of the human adventure in space, we will have to learn about how they are affected also. Thus if *other* space sciences are worth doing, and if a crewed space program greatly advances their cause, then space biology is also justified.

Settling for this argument, however, keeps us from subjecting space biology to the same scrutiny as the other space sciences, thus tacitly accepting the assumption that space biology has little to offer on its own. Now, there are experts, even in biology, who make precisely that assumption, and my immediate task is to examine their reasons.

Part of the problem, as I said earlier, is one of image. For instance, early on, NASA placed great emphasis on vestibular research, since motion sickness is probably the result of vestibular disorientation. And thus people who wanted to turn down support for space biology often said that it was just more research on why astronauts throw up.[40]

But such remarks are neither accurate nor fair. In space we can ask new questions about life. In particular, we can study the role of gravity in the structure and the development of organisms. In a space station, for example, we may choose at will the amount of gravity to which plants and animals will be exposed. This can be done merely by the use of a centrifuge. When the centrifuge is off, the gravity is close to zero. And, when it is on, it makes the container go in circles, subjecting the object under study to different amounts of acceleration. To a plant or an animal such acceleration is equivalent to a gravitational force acting on it.[41] Our main interest lies in the range between 0 and 1 g, so as to study not only the perception of gravity but perhaps even the role that gravity has played in evolution. By experimenting in that range, we may be able to determine gravitational thresholds of biological importance; that is, we may determine the minimum level at which gravity can be detected and at which it becomes a significant factor in physiological or developmental functions.

But are these important questions? Are there any reasons to suspect that gravity will in fact turn out to be a significant biological factor? At first sight, there certainly are reasons. Gravity is all-pervasive in our planet; it is not hard to imagine that life took advantage of its presence to favor some avenues of evolution over others. As Galileo noticed as early as 1638, how much an animal weighs depends on how well its bones can support it.[42] Thus its anatomical structure depends on gravity. In a planet with lower gravity, we may find much taller animals and more symmetrical trees (the symmetry is often broken because slight differences in mass in the branches weigh the tree down in different ways; the more the gravity, the more pronounced those differences become in the development of the tree).

As for the perception of gravity, it influences the way a plant grows. In the microgravity of a space experiment, roots grew out of the ground into the air and the shoots were generally disoriented in spite of the constant illumination from the top. Gravity is obviously a factor in the case of the roots; shoots require both gravity and illumination, although other factors like the distribution of nutrients may play an important part as well.[43] But how do plants recognize gravity? They have gravity receptors, and thanks to space research, we are beginning to understand what those receptors are.[44]

So far there are indications that gravity may also play a role in the axial orientation of amphibian embryos (which is a factor in the normal development of amphibians) and perhaps also in that of birds.[45]

It is not difficult to imagine that what is true of anatomy may also be true of physiology. Indeed, microgravity leads to a shifting of body fluids, and such shifts affect the cardiovascular system in humans. As a result, we have had an opportunity to study how the functioning of the cardiovascular system— or rather its malfunctioning—is connected with the deterioration of muscles. This, of course, is a matter of significance for the general population, especially for the elderly. Moreover, in microgravity we no longer need many of our big muscles to support us. As a consequence, the body begins to reduce its levels of calcium and other minerals needed to strengthen them. This presents a big problem for astronauts, whose bones become weak and brittle. On the other hand, their problem has given us a chance to study the connections between bone and mineral metabolism and endocrine action. Here the adverse reactions of astronauts to weightlessness resemble the symptoms of some diseases on Earth. Space physiology thus offers a chance to investigate the underlying mechanisms.[46] The much-belittled vestibular research, for example, may yield some insights about Meniere's disease, an affliction of the middle ear characterized by deafness and vertigo.

Nevertheless, the value of space biology proper continued to be discounted. The general feeling was that we could do much better if the money, talent, and effort were directed elsewhere in biology. This feeling went hand in hand with the general perception that the biological research done in space was not of very high quality. This low evaluation was based mainly on two concerns. The first was that all the abnormal effects of gravity take place at the systems level, not at the level of cells. Thus, for example, since we do not need strong bones for support, we lose calcium; this loss may in turn have unusual effects on several physiological functions, and so on. But with appropriate exercise and diet we may preserve our strong bones; therefore, the system imbalance will be largely corrected and the unusual circumstances will be kept to a minimum. As a consequence, the biological significance of space will also be kept to a minimum.[47]

The correctness of this notion was presumably buttressed by experimental and theoretical considerations. Most space biologists themselves interpreted the results of many cellular experiments as indications that cells are largely unaffected by gravity.[48] And this conclusion comes as no surprise, since it accords with what theory had led them to expect: cells are small enough that the force of gravity means little when compared to the electromagnetic forces so crucial to the chemical bonds of life.

What Matters in Biology

The second and complementary concern was the notion that what really matters in biology takes place at the molecular level or close to it. Therefore, controlling gravity as a factor was not going to bring us great breakthroughs.

Let me examine these concerns. The first is that the quality of the initial research was not very high. Now, I must admit that there was a clear sense in which this charge was correct: Space biology proper, mostly applied research, initially did not produce much research that qualified as extremely important. Its aim was to gather the preliminary information that could then serve as the inspiration for hypotheses or as the ground for the testing of ideas. Since, except for the clinical research done for the safety of astronauts, most space biology took a back seat to other science, it is not surprising that the information obtained was generally inconclusive and sketchy.

Let me illustrate the nature of the problem by discussing the study of mammalian development. Ideally, we would want to determine the role of gravity by observing whether any of the important stages in development is affected in microgravity. We must look at copulation, fertilization, initial cleavages, embryonic and fetal stages, and postnatal maturation. Many of these are, of course, divided into several distinct and important stages. But we have never been able to observe in microgravity mammalian copulation, let alone post-natal maturation.[49]

The situation is better, but not all that much better, with other animals or plants. Fertilized eggs of fish have flown in space.[50] These eggs have hatched successfully, whereas frog eggs have not produced tadpoles. And plants have been made to produce seeds.[51] Unfortunately yields have been low and chro-mosomal abnormalities common. Some of these results could be accounted for by the stress of flight itself (e.g., the accelerations of take-off and re-entry) or by shortcomings of the life-support systems that make up their artificial space environments. In addition to all that, specimens have normally been examined after their return to Earth, when gravity has begun to reverse the effects of its absence. We need to go from seed to seed and from egg to egg— *in space*. And we need to do this several times over, so that we can isolate and control factors that belong to the inconveniences of spaceflight other than microgravity. But these multigenerational studies would take months, if not years. And then we also need trained biologists to monitor and ex-amine the specimens in space, where the effects we want to discover take place. Living things must be handled delicately and skillfully if they are to bare their secrets. In more recent flights, the reproductive systems of rats have been damaged, with the ovaries shutting down in females and the testes in males shrinking.

The difference between what needed to be done and what was avail-able made space biology appear primitive. To a critic, the space biologist's collections of data resembled the wasteland of Baconian science—without theoretical direction, and thus without theoretical interest.

The critic burdened them with a catch-22, space biologists felt. To show how their field may be significant, they had to carry out enough investigations so they could begin to ask fruitful questions. But critics objected to those investigations on the grounds that space biology had not yet been shown to be significant.

It may be useful to compare the situation of space biology to that of atomic physics at the beginning of the twentieth century and to that of particle physics during the 1960s: physicists would accelerate particles, crash them against targets, and analyze the debris created in the collision (in atomic physics the aim was to determine the structure of atoms by the deflections of the electrons that crashed against them; in particle physics, to discover particles and determine their properties). There is a sense in which the particle and atomic physicists were just "fishing," as the space biologists are now accused of doing.

Critics will no doubt rush to argue that there is a big difference. The difference is presumably that they were fishing in fundamental waters. Whatever they found would have the most significant consequences. But that is easy to say in hindsight. The proliferation of particles in the 1960s led many to think that further searches—and extremely expensive searches at that—would amount to just looking around without any theoretical purpose of note, a Baconian end to a century of exciting physics. As for Rutherford and the other atomic physicists of the turn of the previous century, they were fishing in domains that most other physicists would not even agree were real, let alone significant.[52]

Hindsight is, of course, a wonderful sense. Today we can see that all those particle surveys in the 1960s allowed physicists to classify the particles into families—a classification not unlike that of the chemical table of elements—and then to propose new theories to explain that classification. Thus, the quark hypothesis and a new era of physics were born.

In both cases, however, it took great vision to see the promise of the research. Some important physicists did think that Rutherford was onto something. They held the notion that fundamental aspects of nature might be explained by discovering how matter was put together at the most elementary level. Many decades later, after discovering so many elementary particles, physicists generally thought that the problem was how to make sense of the array. And thus what might have looked like hack work from one point of view, from another looked like the preliminary taxonomy essential for the physics of the future.

My point is precisely that space biology should have been given the chance to carry out that preliminary taxonomy, even if it looked like hack work to some critics. The warrant for doing so is the same as in the two cases from the history of physics: the theoretical payoff comes from testing matter well beyond the range that we have examined previously.

This warrant is clearly seen in the case of some sciences. Take space astronomy, for example. Once we could detect the entire electromagnetic spectrum by placing our telescopes and other instruments above the atmosphere, we embarked in a new survey of the heavens. It may have seemed that we are just looking around. But we had good reason to suspect that what we would find would be significant, that our ideas would be challenged to the utmost.

And that reason was precisely that we knew how limited our range of observation had been until then.[53]

Gravity and Life

In space biology, unlike space astronomy today, but like space astronomy not long ago, we cannot specify the great scientific rewards that await us. We know, however, that the gravity of the Earth has been a constant throughout the evolution of life. Its very pervasiveness, however, makes it difficult for us to determine its role. Surely we cannot resolve the matter by further standard biological analysis. For in analysis we use the tools of prevailing theory to investigate some phenomena, but nothing in our previous biology makes gravity a crucial element of the theory. What we need is either an alternative theory in which gravity is assigned a specific role, or else the manipulation of gravity to make present theory fail. We can do both.

These new theoretical and experimental directions are suggested in part by some of the early results of space biology, and in part by emphasizing some relevant aspects of standard theoretical biology. They permit us to overcome also the second concern that devalued space biology: that microgravity affects organisms only at the systems level. Even if true, it takes away nothing from the promise of space biology proper. If we wish to determine the role of gravity as an all-pervasive factor in individual development and in patterns of evolution, the systems level is actually not a bad place to begin. The human body, for instance, appears fine-tuned for the Earth's gravity. We might have expected to extrapolate our centrifuge studies here on Earth (with gravities above 1 g) to the microgravity of space. Thus, if we lose body mass under an acceleration of several gs, have it normal at 1 g, then presumably we would gain mass at less than 1 g. But it turns out that we lose mass at less than 1 g.[54] Similar reactions take place in microgravity with temperature control and other physiological functions.[55]

This fine-tuning of physiology to the Earth's gravity should provide a fruitful theoretical perspective to study the relationships between a variety of internal systems and cycles in the human body. Why are physiological functions maximized at 1 g? This leads to questions about why the body works as it does, questions that would not occur that easily otherwise. One possible answer is that gravity is used to harmonize a variety of physiological systems—gravity is like a glue that helps hold such systems together. Once the glue is gone, they do not quite work together. And from their failure we learn what makes them work correctly under terrestrial conditions. Another answer is that those systems change their responses in order to adapt to the new conditions. This may also give us significant clues about their normal modes of interaction with other systems or mechanisms.

The fine-tuning to 1 g may become acute in issues of development. In microgravity, a human male excretes from 1.5 to 2 liters of body fluids, with pronounced reductions in the levels of sodium and potassium. By contrast, a pregnant human female is expected, in 1 g, to show an increase of 1.5 to 4 liters over her pregnancy, with a marked retention of sodium. Since the development of the fetus follows a strict sequence in which each event must take place within a critical period, and since the availability and composition of the body fluids are essential to the proper environment in the placenta, we can readily see that disruptions at the system level affect physiological processes at lower levels. At the present time it would be morally impermissible to have pregnant women in the microgravity of space. Laura S. Woodmansee agrees with this conclusion.[56] Further research will be needed to determine whether similar dangers exist in Mars and other places where gravity is lower than on Earth. For social and political considerations about human reproduction in such cases, the reader may consult Konrad Szocik.[57]

The human body is resourceful; it may be able to compensate for the effects of microgravity in a systematic fashion even during a pregnancy. But to determine whether it can, we must resort to experiments on animals.[58] By removing gravity, then, we can observe how the development comes unglued; by "dialing" several degrees of gravity we can refine our examination. The mere fact that many systems function optimally at 1 g provides warrant for designing experiments to determine how the timing and feedback controls of development operate. This, it seems to me, is not a matter of small importance.

"Mere" systemic effects can have profound repercussions that may extend to the cellular level. Changes in the environment of the cell lead to cellular changes in shape, in ability to move, and in internal metabolism (i.e., polarity, secretion, hormone regulation, membrane flow, and energy balance).[59] Changes that take place at the systems level in the organism, such as body fluid shifts, are then bound to affect several cellular systems, change the cellular environment, and thus affect the cells themselves.

The experiments that initially indicated that gravity was irrelevant at the cellular level were performed in cell cultures; they did not examine cells that formed part of the complex wholes that are the cells' normal environments. It is not surprising, then, that such experiments could not expose the indirect action of gravity that starts at the systems level of the organism and works its way down into the realm of the small.

Since genes contain the "language of life," and since organisms are presumably the books written in that language, we tend to think that significance in biology goes from the small to the big. But we have just seen that much of the small depends on the big, since the function of the small depends on the larger whole to which it belongs (and that whole often depends on an even larger one of which it is a part, and so on). It is true, however, that genes are in some sense supposed to be independent of the organism's environment: theory demands that genetic variation not be coupled to the mechanism of selection (i.e., that the environment cannot have a hand in inducing the variations that are compatible with it; otherwise it would be possible to inherit acquired characteristics). Nonetheless, even if this demand is strictly interpreted,[60] many molecular processes may still be open to influence from above.

It is also true that, in many areas of biology, real and fundamental progress is achieved when a strong connection can finally be made to the genome of the organism; that is, when we can finally explain how the genes give rise to the mechanism or function in question. Nevertheless, it is misleading to think of the genome as the blueprint of what the organism will grow into, barring acts of God and other misfortunes. Given a certain genome—and the right circumstances at many different stages of development—a certain individual organism will likely be the outcome. But at many critical junctures within those stages of development, things could go slightly differently. The outcome would be a different organism. It is at those critical junctures that genes and the various structures to which they give rise take advantage of many environmental constants in order to keep their appointed rounds. Nature does not create everything anew and at once. It uses what is already there; it builds on the structures it finds in place; it develops not by reaching for an ideal but by a process best described as jury-rigging. This is true of evolution, but to some extent it describes development as well, particularly the expression of complex genes.

My suggestion is, of course, that the fine-tuning of several physiological functions for 1 g indicates that gravity is one of those constants that provide the context in which the language of life comes to make sense. Without it, aspects of the human genotype would be expressed very differently, if they could be expressed at all. Space biology may thus contribute to an understanding between genetics and the rest of biology. Besides, it is not easy to discover genes and then ask what they do. Often the question goes from the top down: given a certain function or structure, how do genes contribute to bring it about? This indicates that we need knowledge of all the other levels in order to guide genetics. Moreover, it is not necessary that all the alignments and states of equilibrium that many physiological systems reach be encoded in any one set of genes. Some genes may have been selected for because they lead to the construction of an organ or function that finds accommodation with earlier organs or functions. And the genes that lead to those earlier organs or functions are now preserved because the new arrangements are advantageous to the organism as a whole.

The point is this: even if we knew what genes brought about the newer organ or function, and even if we knew all the steps in the construction, we still would not understand that aspect of physiology. For those genes and those steps make biological sense only against the background of the existence of those other organs or functions. And all these together make sense only in the context of whatever constants a form of life has come to take for granted. Thus, for example, the pattern of a net of nerves may not be encoded in the genes. The only "instruction" may be for the nerves to keep growing in search of a particular chemical attractor, but since the tissue in which they grow does not permit easy penetration, the nerves have to work their way around. The result is a very specific pattern, although nowhere in the genome could we find the "blueprint" for such a pattern.[61]

The moral of this story is that knowledge of one level—even if it is "fundamental"—is very limited without knowledge of the other levels. But this more complete knowledge can be gathered only to the extent that we grasp the context in which genes find ultimate expression. And that requires the manipulation of constants in order to determine their role.[62]

Important Discoveries

In any event, the "applied" biology we have done in space has yielded a copious treasure of medical improvements here on Earth, even if operating with a hand tied behind its back. For example, by 1990, it had already led to technological breakthroughs in (1) diagnosis, treatment, and prevention of cardiovascular diseases; (2) new approaches to the understanding of osteoporosis; (3) early detection of genetic birth defects; (4) emergency medical care; and (5) treatment of chronic metabolic disorders.[63]

Although my assessment of space biology was first developed a number of years ago, a very similar assessment can be found in a 2020 paper published by Shelhamer et al.[64] The authors also discuss several discoveries of fundamental scientific and medical importance that have resulted from biological research in space. They begin with techniques to increase bone density in cases of rapid bone loss caused by disuse such as amputations, ligament tears, and lower-leg fractures. They are now used in hospitals around the world. Spaceflight research has also improved our understanding of the connection between diet and bone, including the effectiveness of vitamin D supplements. This research offers the potential to generate further clinical research on the therapeutic and detrimental effects on bone by a variety of nutrients. Thanks to the study of shifting fluids in microgravity, we have learned important lessons about central venous pressure and left ventricular end-diastolic volume. Moreover, we now have knowledge of how gravity and posture influence heart-lung interactions. We have also discovered that gravity may deform old lungs, leading to impaired gas exchange in the lungs of the elderly. Space biology has also led to a better understanding of neuroplasticity in sensorimotor and neuro-vestibular function. For references to the many relevant studies, I recommend consulting Shelhamer's paper, which in example after example reinforces remarks made in this book about the serendipity of space science.

Before moving on to the next topic, I would like to mention one more line of research. In one experiment, samples of salmonella bacteria flew in the Space Shuttle while control samples under identical conditions, except for the presence of gravity, remained on the ground. The space-traveling salmonella became much more virulent because certain genes changed their level of activity in microgravity—genes crucial to the control of a protein called Hfq, which allows bacteria to adapt to changing conditions.[65] This finding, worthwhile in itself, may also help us understand some causes of increased virulence of such bacteria on Earth.

I have argued for the relevance of gravity to biology. But how much gravity is needed for acceptable physiological performance? It may turn out, for example, that the actual threshold for the proper functioning of physiological and developmental is less than 1 g; perhaps as soon as a gravity vector (direction and intensity of gravity) is detected, the plant or animal in question will work normally or almost normally. Determining that threshold should be a goal of the subdivision of the field called "gravitational biology."[66] Indeed, it seems that plants (at least small plants) do the necessary sensing of gravity between 0.1 g and 0.3 g. With animals this sort of research has yet to make much progress. The animal centrifuge in the ISS cannot hold anything bigger than a mouse. These and other results, as well as present and future plans, can be found in the "International roadmap for artificial gravity research."[67]

Two further considerations should give solid anchor to biological research in space. The first is conceptual; the second is historical. H. A. Lowenstam and Lynn Margulis have argued convincingly that the ability by eukaryote cells to modulate their internal concentrations of calcium made possible the appearance of calcareous skeletons.[68] The appearance of such skeletons in turn made possible, over 530 million years ago, the Cambrian Explosion, during which all the major divisions of life became established. The importance of this argument, for our purposes, is that calcium metabolism is one of the first things affected by microgravity.

Calcium ions in solution are essential for many physiological functions, including cell adhesion, muscle contraction, amoeboid cell movement, and they are also used to transmit information between cells. Not only is the intracellular regulation of calcium essential to the function of eukaryote cells, it is also important in the genesis of tissues and embryos. The regulation of calcium seems to be one of those physiological functions on which nature has built a whole array of other functions. This key evolutionary role apparently began, according to Lowenstam and Margulis, when "prey, forced to escape from more effective predators, developed highly integrated sensory and motor systems that must have involved increased coordination and speed of the muscle system."[69] These two skills depend on muscle contraction. And muscle contraction "responds directly to calcium release."[70]

Since the regulation of calcium may have been very instrumental in the evolution of complex organisms, and since life develops in a jury-rigging fashion, the biological importance of calcium may go far beyond what standard theory assigns to it. Determining that importance is precisely one of the areas where experimentation in space can be of advantage. This consideration thus provides one more reason for thinking that in space biology, too, exposure to new circumstances may lead to the profound transformation of our ideas.

The examination of the role of gravity in living things has paid off from the beginning. It was Charles Darwin himself who first noticed that the tips of growing roots and shoots were used to detect gravity. "We now know," he said, "that it is the tip alone which is acted on, and that this part transmits some influence to adjoining parts, causing the latter to bend."[71] The study of this influence, in a long series of experiments beginning with the publication of Darwin's work in 1880, yielded the separation (1920s) and chemical identification (1930s) of the first hormone known to make plants grow.[72] Space biology proper has seized the opportunity to continue this history.

Early on, though, space biologists had to respond to the question whether the large expenditures of space biology were justified when its significance was more of promise than of record. "$50,000 can pay for a decent experiment in standard biology today," a biologist told me in the 1980s. But for that money we could hardly get a rabbit into orbit, to say nothing of enough rabbits for biologists to supervise multigenerational studies in space. We must notice, however, that this complaint did not deny that space may be beneficial to biology, or that it may be beneficial in important ways. Its aim was rather to make us favor Earth-based investigations—which we already knew were very important—over the expensive biological experimentation in space.

Space biologists often responded in two ways. First, they pointed out that their budgets were small as space budgets went (in the 1980s, for example, it was about $30 million a year—including the monies for astrobiology, gravitational biology, and flight experiments—which was roughly about 3% of the total space science budget, and not much when compared to all the monies spent for biological research on Earth). Their second point was that most of the money for space biology was not going to come from the general funds earmarked for the life sciences. The money was going to come from space exploration funds.

Critics were unlikely to accept these two points. First of all, even if the space biology budget was a pittance, to do the research properly would take far greater amounts. For example, if space biology was used as one of the justifications for the International Space Station, we would have to think in terms of hundreds of millions of dollars. As for the second point, to say that the two were not really in competition for the same monies was an easy out. For whatever the name of the account, in the long run society pays for both. Why should space biology take so large a share?

Nevertheless, the matter was not as simple as that. Space biology does indeed belong to a large space commitment. Because of that, the notion of competition between standard biology and space biology flounders. To see why, imagine what happens to a man who buys an automobile and wants a radio in it. For the amount of money that he will have to spend, he could have gotten several radios far superior to the one that goes into his car. But those radios would not do quite the same job effectively. Similarly, since we are going into space, we have an opportunity to investigate life in ways not open to us before. The whole question of doing space biology must then be taken in the context of having a space program, just as the question of having the car radio arises in the context of having a car and not in that of purchasing radios generally. All the while we must keep in mind how promising a radio this is.

A crucial qualification: microgravity clearly is inimical to the health of the astronauts in long missions, and thus artificial gravity will have to be developed; otherwise such missions will be jeopardized. Once that technology is in place, microgravity experiments on plants and animals will still be possible in unattached modules, for example, but the long-term exposure of humans to such conditions may be deemed contrary to the ethics of medical research. Today's astronauts give their fully informed consent, knowing well that those risks are largely unavoidable, and rather small by comparison with the danger of not coming back alive at all. Explorers have often risked their lives for the chance to participate in a great adventure. But as technology advances, some of those risks become unacceptable. No one wishes, anymore, to put his life on the line for the opportunity to visit the North Pole. Those who first went there, however, had to be willing to pay that price.

Nevertheless, we need to tread carefully here, for we do have an ethical obligation to the safety of astronauts. A most important issue in this regard is the potential use of biological enhancements in the human exploration of space, particularly in the exploration of Mars. The argument I will present can be extended to other instances of long-term space exploration.

There are two main kinds of health threats involved in the exploration of Mars. One is the threat from long exposure to microgravity during a trip that lasts about seven months in each direction. The other comes from exposure to radiation both during the trip and on the surface of Mars. Let me begin with the notion of changing the genetic endowment of explorers so as to make them immune to radiation. This dream seems closer to realization thanks to a new technology, CRISPR-Cas9, which will presumably lead to the genetic cures of a multitude of health problems. The idea is then to turn on genes or to insert them (from other species) that would give protection against radiation. Unfortunately, a new and well-regarded study shows that, far from being specific in its application, the deletion technology often resolves into deletions found over many thousands of bases.[73] Crossover events and lesions distal to the cut site seem common as well. This genomic damage may also lead to cancer and other pathogenic consequences.

Taking such risks while deleting or modifying defective genes could be justified ethically when the patient's life is in the balance. But in the case of the explorers, we would be deleting or modifying *nondefective* genes. More important, during the trip to Mars, existing technology can prevent radiation damage from solar flares and the like. The real danger comes from cosmic rays, which are protons, helium, and other particles traveling close to the speed of light. The radiation they produce when hitting the walls of the spacecraft can, nonetheless, be stopped by the hydrogen in water and in plastics such as polyethylene. From now on, we could put the water supplies in polyethylene containers next to the outer walls, for example. In addition, NASA is developing hydrogenated boron nitride nanotubes (hydrogenated BNNTs), the ideal shield against radiation, both for spaceships but also for dwellings and vehicles on Mars. Moreover, this material can be made into yarn that will clothe the explorers themselves. Given this technological alternative, genetic interventions on astronauts cannot be ethically justified.

Let me address now medical interventions to reduce the damage of long exposure to microgravity during the trip to Mars. Implants have been tried on mice to control muscle atrophy in microgravity.[74] The goal is to transfer the technology to humans. As a general approach to dealing with the problems caused by absence of gravity, however, we would have to then use different implants to deal in addition with osteoporosis, problems with the transport of fluids, as well as effects on many types of cells, the immune system included.[75] Therefore we would overload the explorer's body with implant systems, to say nothing of causing the potentially dangerous interactions of multiple types of drugs.

In the meantime, nonmedical solutions to the problem of long exposure to microgravity have been discussed for over a century in scientific and popular writings,[76] and particularly artificial gravity created by rotating structures.[77] My own version would have a spaceship in which the different components (rocket, supply compartment, and human compartment) instead of being stacked up (with the rocket in the bottom, obviously) are assembled side by side. The rocket would be in the middle and the other two compartments connected to the rocket by a cable or thin structure, hundreds of meters apart from each other, and made to rotate (not difficult). Astronauts, inside their compartment as it rotates, are subjected to acceleration comparable to Earth's, or less if the research discussed earlier so indicates.

Let us then carry out space biology in more insightful ways, but in doing so let us take pains to preserve our sense of humanity.

Notes

1. The following research of mine was helpful in the first section of this chapter: "Science and Ethics in the Exploration of Mars," in *The Human Factor in a Mission to Mars: An Interdisciplinary Approach*, ed. Konrad Szocik (Springer, 2019), pp. 185–200; "Ethical Obligations Towards Extraterrestrial Life," *Philosophy Study* 10, no. 3 (March 2020): 193–201. doi:10.17265/2159-5313/2020.03.003.

2. I do not divide the field in the manner that NASA has found convenient for a variety of administrative reasons (according to which, for example, Space Biology is only a small section of the Life Sciences Division). My division follows rather the convenience of the argument and of the reader.

3. Yan Zhou, Xuexia Xu, Yifeng Wei, Yu Cheng, Yu Guo, Ivan Khudyakov, Fuli Liu, Ping He, Zhangyue Song, Zhi Li, Yan Gao, Ee Lui Ang, Huimin Zhao, Yan Zhang, and Suwen Zhao, "A Widespread Pathway for Substitution of Adenine by Diaminopurine in Phage Genomes," *Science* (April 30, 2021): 512–516.

4. Adapted from R. M. Laws, I. S. C. Parker, and R. C. B. Johnstone, *Elephants and Their Habitats* (Clarendon Press, 1975).

5. For an informal, though detailed account of these issues, see Gene Bylinsky, *Life in Darwin's Universe* (Doubleday, 1981).

6. For a very accessible account of the classic Miller-Urey experiment, see D. Goldsmith and T. Owen, *The Search for Life in the Universe* (Benjamin/Cummings, 1980), pp. 174–177.

7. Martin Ferus, Fabio Pietrucci, Antonino Marco Saitta, Antonín Knížek, Petr Kubelík, Ondřej Ivanek, Violetta Shestivska, and Svatopluk Civiš, "Nucleobases in a Miller–Urey Experiment," *Proceedings of the National Academy of Sciences* 114 , no. 17 (April 2017): 4306–4311. doi:10.1073/pnas.1700010114.

8. Several alternative liquids have been proposed, especially ammonia and methyl alcohol. Goldsmith and Owen offer a very accessible account of this matter also (*The Search for Life in the Universe*, pp. 212–216). For more technical literature, the reader may consult the appropriate titles in note 8.

9. L. Margulis and J. E. Lovelock, "Atmospheres and Evolution," in *Life in the Universe*, ed. John Billingham (NASA cp.2156, 1981), p. 79.

10. For this point of view, see p. 5024. S. M. Siegel, "Experimental Biology of Extreme Environments and Its Significance for Space Bioscience, 1 and 2," in *Spaceflight*, 12: 128–130; 256–299 p. 128; Siegel et al., "Experimental Biology of Ammonia-Rich Environments: Optical and Isotopic Evidence for Vital Activity in Pennicillium in Liquid Ammonia-Glycerol Media at –40 C," *Proceedings of the National Academy*

of Sciences 60, no. 2 (1968): 505; S. M. Siegel and T. W. Spettel, "Life and the Outer Planets: II. Enzyme Activity in Ammonia-Water Systems and Other Exotic Media at Various Temperatures," *Life Science and Space Research* 15 (1977): 76; B. Z. Siegel and S.M. Siegel, "Further Studies on the Environmental Capabilities of Fungi: Interactions of Salinity, Ultraviolet Irradiation, and Temperature in Penicillium," in *Gospar Life Sciences and Space Research*, ed. R. Holmquist, vol. 8 (Pergamon Press, 1980), p. 59.

11. A. Affholder, F. Guyot, B. Sauterey, R. Ferrière, and S. Mazevet, "Bayesian Analysis of Enceladus's Plume Data to Assess Methanogenesis," *Nature Astronomy* 5 (2021): 805–814. https://doi.org/10.1038/s41550-021-01372-6

12. Other rocky moons with possible oceans include the gigantic Jovian moon Ganymede: R. Cowen, "Ganymede May Have Vast Hidden Ocean," *Science News* 158 (December 23 & 30, 2000), p. 404.

13. https://www.esa.int/Science_Exploration/Space_Science/Mars_Express/Signs_of_ancient_flowing_water_on_Mars2

14. *Science* on radial shift as cause of dendritic channels.

15. "Special Section: Mars Reconnaissance Orbiter," *Science* 317 (September 21, 2007): 1705–1719.

16. M. Carr, "The Proof Is in: Ancient Water on Mars," *The Planetary Report* XXIV, no. 3 (May/June, 2004): 11.

17. R. Orosei et al., "Radar Evidence of Subglacial Liquid Water on Mars," *Science* 361, no. 6401(2018): 490–493. doi:10.1126/science.aar7268.

18. D. S. McKay et al., "Search for Past Life on Mars: Possible Relic Biogenic Activity in Martian Meteorite AL84001," *Science* 273 (August 16, 1996): 924–929.

19. See R.A. Kerr, "Ancient Life on Mars," *Science* (1996):864–866. The controversy spread to public arguments in the newspapers; see, for example, the front-page article "Life on Mars: Scientists 'Thrilled' by Prospect," *Seattle Times*, August 7, 1996.

20. J. Martel and J. Ding-E Young, "Purported Nanobacteria in Human Blood as Calcium Carbonate Nanoparticles," *Proceedings of the National Academy of Sciences* 105, no. 14 (2008): 5549–5554.

21. B. Luef, K. R. Frischkorn, K. C. Wrighton, H. N. Holman, G. Birarda, B. C. Thomas, A. Singh, K. H. Williams, C. E. Siegerist, S. G. Tringe, K. H. Downing, L. R. Comolli, and J. F. Banfield, "Diverse Uncultivated Ultra-Small Bacterial Cells in Groundwater," *Nature Communications* (2015), 10.1038/ncomms7372. This research was helpful in the question of ethical obligations in astrobiological exploration.

22. Although the emphasis has been mine all along, in making these remarks, I find myself paraphrasing Harold P. Kline's many comments on earlier drafts of this essay.

23. S. J. Mojzis et al., "Evidence for Life on Earth before 3,800 Million Years Ago," *Nature* 384, no. 6604 (1996): 55–59.

24. Kevin A. Maher and David J. Stevenson, "Impact Frustration of the Origin of Life," *Nature* 331, no. 6157 (1988): 612–614.

25. V. S. Johnston, *Why We Feel* (Perseus Books, 1999), p. 44.

26. David N. Reznik, "From Life to Death: The Evolution of Life Histories in Guppies," in *Sex, Reproduction and Darwinism*, ed. Filomena de Sousa and Gonzalo Munévar (Pickering and Chatto, 2012), pp. 9–32.

27. A. Lazcano and S. L. Miller, "How Long Did It Take for Life to Begin and Evolve to Cyanobacteria?" *Journal of Molecular Evolution* 39, no. 6 (1994): 546–554.

28. C. de Duve, *Life Evolving* (Oxford University Press, 2002); C. de Duve, *Singularities* (Cambridge University Press, 2005).

29. Alexander Graham Cairns-Smith, *Seven Clues to the Origin of Life* (Cambridge University Press, 1990; Canto reprint of the original 1986 edition).

30. R. P. Turco, O. B. Toon, T. P. Ackerman, J. B. Pollack, and C. Sagan, "Nuclear Winter: Global Consequences of Multiple Nuclear Explosions," *Science* 222 (December 23, 1983): 1293. See also O. B. Toon et al., "Evolution of an Impact-Generated Dust Cloud and its Effects on the Atmosphere," *Geological Society of America* 190 (1982): 187.

31. See, for example, R. F. Service, "Researchers Create First Autonomous Synthetic Life Form," *Science* 299 (January 31, 2003): 640.

32. Hoyle's views on this issue and panspermia in general can be found in his *Evolution from Space* (Simon and Schuster, 1981), and *The Intelligent Universe* (Holt, Rinehart and Winston, 1984).

33. R. Shapiro, *Origins: A Skeptic's Guide to the Creation of Life on Earth* (Bantam Books, 1987).

34. For a fuller treatment of this issue, see Chapter 8.

35. S. M. Marshall, C. Mathis, E. Carrick, et al., "Identifying Molecules as Biosignatures with Assembly Theory and Mass Spectrometry," *Nature Communications* 12, no. 3033 (2021). https://doi.org/10.1038/s41467-021-23258-x

36. For an alternative approach, see A. Azua-Bustos and C. Vega-Martínez, "The Potential for Detecting 'Life as We Don't Know It' by Fractal Complexity Analysis," *International Journal of Astrobiology* 12, no. 4 (2013): 314–320. doi:10.1017/S1473550413000177.

37. *Evaluation of Bioburden Requirements for Mars Missions* (2021). Available at nap.edu/26336.

38. James S. J. Schwartz, *The Value of Science in Space Exploration* (Oxford University Press, 2020).

39. This emphasis is exemplified by the attitude of the National Academy of Sciences. In a report entitled "Space Station Needs and Characteristics" (May 1983), the Committee on Space Biology and Medicine of the Academy's Space Science Board said, "the main scientific justification for a Space Station Biomedical Laboratory is laying the physiological groundwork necessary for launching crewed space flights of long duration sometime in the next century. . . . Although [the zero-g] environment would provide also an opportunity for carrying out some fundamental biological research, we do not believe that this aspect can be a major consideration in justifying the Station" (p. 4). The fundamental biological problems recognized by the academy are the perception of the gravitational vector by plants and the determination of body axes in metazoan development (chiefly in amphibians and birds—both problems are discussed in this chapter). It seems that just two problems are not enough, or perhaps these two are not fundamental enough, in the eyes of the academy.

40. Scientists who oppose crewed exploration on the grounds that it detracts from real space science often concentrate their fire on space biology. Writing in *Nature*, R. Jastrow said that the Space Station would be a tragedy, "another two decades of

original research on why astronauts vomit" (quoted in *Science Digest*, May 1984, p. 142).

41. The main difference is that Coriolis forces may be more pronounced in centrifuges.

42. A good source for the 1980s state-of-the-art research on how gravity affects life can be found in the proceedings of the annual meetings of the IUPS Commission on Gravitational Physiology, published as supplements to *The Physiologist*. See particularly 25, no. 6 (Dec. 1982) and 27, no. 6 (1984). For biology in the Space Shuttle see the series of reports from Spacelab entitled "Life Sciences," *Science* 225 (July 13, 1984): 205–234. For possible future experimentation, see *The Fabricant Report on Life Sciences Experiments for a Space Station*, ed. J. D. Fabricant, a publication of the University of Texas Medical Branch, Galveston, Texas, 1983.

43. P. Zabel, M. Bamsey, D. Schubert, and M. Tajmar, "Review and Analysis of over 40 Years of Space Plant Growth Systems," *Life Sciences in Space Research* 10 (2016): 1–16.

44. See the gravitational physiology supplements to *The Physiologist* cited earlier. Also see "Experiments on Plants Grown in Space," Supplement 3 to *Annals of Botany* 54 (Nov. 1984). For an assessment and long-range planning of plant gravitational research, see "Plant Gravitational and Space Research," Report of a Workshop held April 30– May 2, 1984, in Rosslyn, Virginia, a publication of the American Society of Plant Physiology, 1984.

45. See, for example, S. Kochav and H. Eyal-Giladi, "Bilateral Symmetry in Chick Embryo Determination by Gravity," *Science* 171 (1971), p. 1027; and A. W. Neff and G. M. Malacinski, "Reversal of Early Pattern Formation in Inverted Amphibian Eggs," in the *Proceedings of the Fourth Annual Meeting of the IUPS Commission on Gravitational Physiology*, p. 119.

46. Paul C. Rambout, "The Human Element," in *A Meeting with the Universe: Science Discoveries from the Space Program* (NASA, 1981), p. 142. In renal and electrolyte physiology also, space brings about many interesting variations from normal.

47. Although similar research could still be carried out on animals.

48. For this general conclusion about the space environment, see G. R. Taylor, "Cell Biology Experiments Conducted in Space," *BioScience* 27 (1977), p. 102. For an influential experiment on cultures of embryonic lung cells, see P. O'B. Montgomery Jr. et al., "The Response of Single Human Cells to Zero-Gravity," in Biomedical Results from Skylab, ed. R. S. Johnston and L. F. Dietlin (NASA SP-377, 1977), p. 221.

49. From the "Final Report of the Developmental Biology Working Group" (unpublished, extended to me as a courtesy by Dr. Emily Holton), p. 3.

50. For a brief summary, see Taylor, "Cell Biology Experiments Conducted in Space," (1977) p. 103n19. For a full report, see H. W. Scheld et al., "Killifish Development in Zero-G on COSMOS 782" (NASA TM-78525, 1976), p. 179. Unpublished Soviet studies, however, have raised serious doubts, in J. R. Keefe's opinion, "about the ability of normal vertebrate fertilization under spaceflight conditions" ("Gravity Is a Drag," unpublished, p. 6).

51. See the *Annals of Botany* supplement cited in note 17. Dr. Holton has made available to me data from Soviet experiments that show very low yields and that suggest that flowering and fertilization may be sensitive to gravity (as indicated, e.g., by lost

chromosomes in Dwarf Sunflower and broken chromosomes in oats). When cells are studied not in culture but as part of a system, there is a peculiar increase in chromosomal abnormalities in the majority of organisms, including humans. See, for example, L. H. Lockhart, "Cytogenic Studies of Blood (Experiment M111)," *Biomedical Results from Skylab*, p. 217. Investigators are much too quick, in my opinion, to explain away such widespread abnormalities as possible effects, exclusively, of flight stresses.

52. Whether atoms existed at all was a matter of controversy among physicists. Ernst Mach, for example, argued that their existence was not a scientific claim. The controversy was settled in 1905 with the publication of Einstein's work on Brownian motion.

53. We might ask whether it is fair to have to spell out the warrant for investigating the role of gravity in biology. After all, the history of science is full of cases where investigations that were considered preposterous led to the most profound changes in thought. But we cannot take this easy way out when space biology requires large sums of money and its critics have credentials and influence. The issue must be faced.

54. Intracellular mass, not including fluids or calcium loss in the bones.

55. Apparently the shift in body fluids affects the communication between cells.

56. Laura S. Woodmansee, *Sex in Space* (Collector's Guide Publishing, 2006).

57. Konrad Szocik, Rafael Elias Marques, Steven Abood, Aleksandra Kędzior, Kateryna Lysenko-Ryba, and Dobrochna Minich, "Biological and Social Challenges of Human Reproduction in a Long-Term Mars Base," *Futures* 100 (2018): 56–62.

58. Of particular interest would be animals who exhibit a highly differentiated ability to distribute fluids (e.g., Gerbilline rodents). "Final Report of the Developmental Biology Working Group," p. 6.

59. "Final Report of the Developmental Biology Working Group," p. 5.

60. Every system of the organism tries to maintain homeostasis (a relatively stable state of equilibrium) against the next higher level, and the organism as a whole against the environment. Thus the genes benefit from many levels of homeostasis serving as a buffer zone against the environment. Nonetheless it is clear that the general environment sometimes will affect the intracellular environment and thus may conceivably act as an agent of selection against some pieces of DNA and in favor of others (i.e., at the molecular level). This is not to say that acquired characteristics can be passed on, since the features of the environment that will favor some traits at the level of the organism are not of a kind with those that would act on pieces of DNA within the cell. That is what happens, for example, when a change in the cellular environment may permit a mutation to survive which will later be considered a defect of the organism vis. a vis. the general environment.

61. Stent's work on nerves is discussed in Chapter 8.

62. The point I am defending here is a corollary of the views I expressed in Ch. 8 of my *Radical Knowledge: A Philosophical Inquiry into the Nature and Limits of Science* (Hackett, 1981). It is also an expansion of Gunther Stent's work on the "meaning" of the genetic code, also discussed on Chapter 8.

63. V. Garshnek, A. E. Nicogossian, and L. Griffiths, "Earth Benefits from Space Life Sciences," *Acta Astronautica* 21, no. 9 (1990): 673–676. doi:10.1016/0094-5765(90)90079-z.

64. M. Shelhamer, J. Bloomberg, A. LeBlanc, G. K. Prisk, J. Sibonga, S. M. Smith, S. R. Zwart, and P. Norsk, "Selected Discoveries from Human Research in Space That Are Relevant to Human Health on Earth," *NPJ Microgravity* 6 (2020): 5. doi:10.1038/s41526-020-0095-y.

65. This example is taken from S. Williams, "Bugs in Space," *Science News* 172 (September 29, 2007): p. 197.

66. In general, centrifuge studies on Earth can be extrapolated to space conditions only with the greatest of cautions.

67. G. Clément, "International Roadmap for Artificial Gravity Research," *npj Microgravity* 3, no. 29 (2017). https://doi.org/10.1038/s41526-017-0034-8.

68. H. A. Lowenstam and Lynn Margulis, "Evolutionary Prerequisites for Early Phanerozoic Calcareous Skeletons," *BioSystems* 12 (1980), p. 27.

69. Lowenstam and Margulis, "Evolutionary Prerequisites for Early Phanerozoic Calcareous Skeletons," p. 36.

70. Lowenstam and Margulis, "Evolutionary Prerequisites for Early Phanerozoic Calcareous Skeletons." See also S. J. Roux (ed.), *The Regulatory Functions of Calcium and the Potential Role of Calcium in Mediating Gravitational Responses in Cells and Tissues* (NASA CP-2286, 1983).

71. C. Darwin, assisted by F. Darwin, *The Power of Movement in Plants* (John Murray, 1880), p. 592.

72. Adapted from the Dedication to Charles Darwin in the *Proceedings of the Sixth Annual Meeting of the IUPS Commission on Gravitational Biology*, September 18–21, 1984, Lausanne, Switzerland.

73. M. Kosicki, K. Tomberg, and Allan Bradley, "Repair of Double-Strand Breaks Induced by CRISPR–Cas9 Leads to Large Deletions and Complex Rearrangements," *Nature Biotechnology* 36 (2018): 765–771. Also see "Science and Ethics in the Human-Enhanced Exploration of Mars," in *Human Enhancements for Space Missions: Lunar, Martian, and Future Missions to the Outer Planets*, ed. Konrad Szocik (Springer, 2020), pp. 113–124.

74. NASA's Rodent Research-6, 2018.

75. J. Pietsch, J. Bauer, M. Egli, M. Infanger, P. Wise, C. Ulbrich, and D. Grimm, "The Effects of Weightlessness on the Human Organism and Mammalian Cells," *Current Molecular Medicine* 11 (2011): 350–364.

76. K. Tsiolkovsky, "Investigation of Outer Space Rocket Devices," *The Science Review* (1911).

77. G. K. O'Neill, *The High Frontier: Human Colonies in Space* (William Morrow & Company, 1977).

7

Humankind in Outer Space

The physics Nobel Laureate Steven Weinberg claimed that "the whole manned spaceflight program, which is so enormously expensive, has produced nothing of scientific value." This remark capped a scathing critique according to which "The International Space Station is an orbital turkey. . . . No important science has come out of it. I could almost say no science has come out of it."[1] And a *New York Times* columnist reportedly stated matter-of-factly that "Three decades after going to the moon, NASA is sending astronauts a few hundred miles above Earth to conduct high school science experiments."[2] Nor is the value of crewed exploration likely to change, for, as Weinberg told SPACE.com, "Human beings don't serve any useful function in space" . . . "They radiate heat, they're very expensive to keep alive and unlike robotic missions, they have a natural desire to come back, so that anything involving human beings is enormously expensive." This derisive view of crewed exploration has not been unusual among space scientists themselves. Indeed, space scientists objected to projects like the International Space Station (ISS) and President George H. W. Bush's 1989 proposals to send humans to the Moon and eventually to Mars. I agree to this extent: exploration by humans, since the end of the Apollo program, has generally been detrimental to space science. And in the short run it will continue to be so. But in the long run, human-led exploration may help greatly the cause of space science while providing great benefits to the human species.[3]

Human versus Machine Exploration: The Near Future

Some space scientists argue that we can achieve the goals of space science better with machines than with astronauts. Their feeling is that the money and effort that could be spent on driving science and technology to explore the cosmos are in large part lost when we concentrate instead on ensuring the safety of astronauts and on developing very expensive and cumbersome life-support systems. An astronaut needs air, water, food, and protection

The Dimming of Starlight. Gonzalo Munévar, Oxford University Press. © Oxford University Press 2023.
DOI: 10.1093/oso/9780197689912.003.0007

from a hostile environment. The satisfaction of these needs requires bigger rockets to handle the far heavier payloads—as well as far more reliable spacecraft. A human being is a delicate creature. Even with the best of our technology we could not easily send astronauts into the hell of Venus or the intense radiation of Jupiter's vicinity. A trip to Mars would also be very difficult, since it would take months under constant bombardment from radiation and the possible threat of solar flares and deadly cosmic rays, in addition to the adverse physiological effects from such a long exposure to weightlessness. Machines, on the other hand, can go practically everywhere in the solar system, for far less money.

I could point out again that some of these problems have solutions, as we saw in the previous chapter, but then they would not even arise with machines. Now, proponents of human spacecraft reply that humans can do many things that machines cannot. For example, humans can perform experiments that require great dexterity, and they can retrieve and fix satellites. That is true, but according to their critics, not relevant. First, machines can do their more limited job in places where humans cannot or should not do any job at all. This includes not only trips of very long duration and hazardous environments but also dangerous experiments. Second, in many space science experiments, the human presence is a hindrance. For example, telescopes have to be so precisely aimed that someone moving around in the spacecraft would disturb the observations. Third, even if astronauts can retrieve and fix satellites, and build and operate industrial facilities, whereas present machines cannot, we can design our space equipment so robots or teleoperators could handle the job. This, of course, requires that we develop robots and teleoperators equal to the task. Thus, on the whole, the impetus that technology receives is greater from exploring space with machines than from worrying about the safe transportation of astronauts. Furthermore, advances in machine operations in low orbit can be applied throughout the solar system, whereas astronauts will be unlikely to venture beyond Mars in the next forty years, and even that looks to the critics like pie in the sky (pun intended).

The tragic destruction of two Space Shuttles, Challenger and Columbia, threw into disarray the space program and has done great harm to space science. They have clearly shown the risks both to human life and to science from too great a reliance on human exploration. Indeed, even when the Space Shuttle flew normally, space science suffered, and, as we will discuss later, the ISS makes matters worse. Why should we then insist on human exploration when we can accomplish far more, and to do it far more cheaply, safely and efficiently with machines?

Exploring with Machines

Let us take a look at the two main technologies favored by the critics of human exploration: teleoperators and robotics.

Teleoperators

Teleoperators permit us to handle via radio and television tasks that must be carried out at a distance. For example, a television camera on a machine transmits to the ground an image of two building blocks, and a human operator makes the arms in the machine put the two blocks together by radio transmission. Interaction between ground crews and a variety of orbiting observatories has actually become routine. We can change orbits, aim cameras and telescopes, and even perform experiments. Advances in the various aspects of teleoperators—sensors, arms, fingers, grip, and dexterity—will increase the range of activities that we can perform by remote control in space. Teleoperators combined with robotics can go even further: the human operator would perform certain repairs, say, in a comfortable laboratory, while a robot would mimic the same actions in a far more hostile environment.[4]

But some serious problems remain. The most obvious problem is that the farther away the spacecraft is, the harder it is to run it by remote control. Imagine that a roving vehicle on the Moon comes suddenly upon a hole or an unexpected rock. Its television camera will immediately send a picture of the obstacle to an operator on Earth. But on the average it takes that signal one and a half seconds to arrive. If the human operator reacts instantly, the instruction will arrive at the Moon one-and-a-half seconds after that. Anyone who has driven a car knows very well that many disasters can be crammed into three seconds, which is the minimum time that it would take for the earthbound human operator to react to the lunar environment. For a rover on Mars that time would be of the order of eight to forty minutes, and for the outer planets we have to allow hours.

There are two ways to reduce this difficulty. One is to anticipate as much as possible and build our space machines accordingly. We could, for example, provide the lunar vehicle with a computer map of the land it must travel (drawn from photographs taken by orbiting spacecraft). Any deviation from that landscape will automatically force it to stop until it receives fresh instructions from its human operator.

On the Earth, of course, an attentive human driver is able to detect a nasty pothole and get out of harm's way in less than three seconds. But to do so, the driver uses a variety of perceptual clues that allow him or her to spot a hole for what it is and to tell just how far it is. The teleoperator, by contrast, is looking through a television camera at an alien landscape: such remote vision is poorer than the terrestrial driver's, and without all the perceptual clues needed to come to a quick decision. The upshot of all this is that the roving vehicle must move very slowly. That is so even when the terrain is reasonably well known. When the vehicle is called upon to do some honest-to-goodness exploration, the difficulty becomes acute.

As a matter of fact, the Russians sent one such vehicle to the Moon. And NASA had plans for another at one time. But when the best hopes for its performance were so clearly surpassed by the actual performance of the astronauts, the project—called "Prospector"—died of natural causes.

The teleoperator is then at a disadvantage with respect to a human on the spot, at least for certain kinds of jobs. Not only is the camera not as good as a human eye, it does not receive the correcting feedback of the other human senses. Human beings do not just see what is out in the world and correctly report it to consciousness. They pick out and concentrate instead on a variety of subtle clues as to what is most relevant and worthy of attention. A human observer, say Paul, looking through a television camera has fewer of those clues (clues that are peripheral, in the background, or correlated with hearing, smell, or touch). Even if Paul is highly trained, in a new situation his degraded experience may not suffice for him to recognize objects and situations in the way he has been trained to do. A laboratory biologist, say Susan, can tell at a glance that a guinea pig is ill because somehow its behavior differs subtly from patterns to which the biologist responds even if she cannot describe them. It is our hands-on experience that allows us to gain what appears to be an intuitive "feel" for our surroundings. Geologists and materials experts may also depend on the immediacy of contact in order to grasp the object of study in this quasi-intuitive way.

In space, it is true, many of the associations between the senses may be disturbed by the absence of gravity, and thus previous training may lead a scientist on the spot to misjudge the situation. But human beings have the capacity to adapt and to form new associations. A biologist making slides of a rat's brain can learn how to compensate for the new environment. The teleoperator Paul, on the other hand, faces two different problems. First, at present his artificial "hand" simply does not have the dexterity to carry out that refined a task. Second, even if it did, and eventually it might, he could not use that artificial hand the way he would on Earth; he would have to be retrained so as to make that artificial hand do from Earth what the biologist does with her real hand in space.

To make a better artificial hand (or other appropriate tools), it would be wise to observe what the human biologist does in her space laboratory, and then slowly refine the technology until it is acceptable. Cutting a rat's brain in weightlessness may be quite different from doing it on the surface of the Earth. It makes sense, then, first to try to learn what it is like to do that kind of lab work in space. That is, we need to develop a space expertise in those activities—a human expertise. Only then we might be in a position to calibrate our teleoperations. Otherwise teleoperators are likely to reduce the human ability to make discoveries and to select what may be of scientific importance.

The obvious conclusion of all this is that teleoperating technology is best developed in cooperation with human activity in the relevant areas. To perform at the level that the proponents of uncrewed flight hope for, we would require to have a joint, and to some degree a prior human presence in space. This might be a useful technology to develop in the ISS. Although a space station is not very useful to all branches of space science, it may be very valuable to biology, materials processing, and medical technology. Moreover, as our experience with the Shuttle has demonstrated, there are experiments in physics that require a high degree of finesse or complexity in the handling from space (e.g., electron beam experiments to examine the interaction of charged particles with the Earth's magnetic field). Without the mission and payload specialists in the spacecraft, those experiments could not have taken place for decades, if at all.

For materials processing, one of the most promising areas of space industrialization, artistry is as crucial as scientific craft. We must not forget that the initial purpose of the ISS, apart from science, is to do industrial research rather than to set up actual industries. If it were the latter, then there might be some hope for combining teleoperators and a high level of automation. But insofar as the purpose is largely one of exploration, the machine technology is not yet up to the challenge. Nor are our teleoperating abilities so developed that we could build those completely automated factories without the help of human workers in space.

Robotics

The other hope of the opponents of crewed spaceflight is the development of intelligent machines. As those critics constantly remind us, computers can already perform better than humans in several areas. One such area is in geology itself—apparently in contradiction to my earlier remark—where expert programs do a better job in the exploration of underground oil deposits. All we need to do is develop expert programs about the solar system, and we will produce robot spacecraft that can do the exploration for us. It may be more heroic to do it with humans, but, according to the critics, at those prices we can afford a bit less heroism and a lot more common sense. Some NASA officials fear that machine-only programs will not enjoy as much public support. In their view, the public derives great satisfaction from the vicarious participation in the grand adventure of space exploration. But opponents of human exploration argue that robots will permit vicarious participation by the general public in the unraveling of the mysteries of the solar system. We have already experienced that pleasure with the Viking landings and the recent rovers on Mars, with the Voyager missions to Jupiter and Saturn, the Galileo spacecraft, and more recently with Cassini's visit to Saturn and Titan.

There is no question that our exploration of the solar system will be more fruitful the more independent and flexible our spacecraft become. But we should not turn this reasonable wish into wishful thinking. The truly successful expert programs are very few; and the successful ones have very limited applications. The spacecraft cannot take expert programs for every contingency unless the mission is rather simple and we have a fair idea of what those few contingencies may be. An expert program, whether for diagnosing diseases or finding oil, works by compiling a large set of techniques and rules of thumb used by the human experts in the field. The programmer discovers what rankings and relative values those human experts give to those techniques and rules of thumb in typical cases of application. Such a program can thus do a better job than an individual human expert because the computer can keep track of a larger number of considerations.[5] But when the situation is not typical, when it demands different rankings and values, as is normally the case with the scientific exploration of the unknown, then the expert program soon exhausts its usefulness. Nor can we program in advance those different assignments of rankings and values, precisely because their appropriateness will be determined by unknown circumstances. This is not to say that human beings seldom make mistakes in the way they grasp the situation. Nonetheless, their flexibility does give them an edge where they meet an open-ended environment.

This problem of flexibility is perhaps the greatest barrier to artificial intelligence. There are many programs that perform very well in restricted domains, but no one has an inkling of how to make a program of general application. All too often what from a distance seems to be a difference in degree that can be overcome with larger computing power and memory storage, up close becomes an insuperable difference in quality. The things that computers cannot do are those like using language or going shopping that come so naturally for even the dullest of human beings. Live intelligence constructs a world for itself (i.e., "interprets" the world as it interacts with it). But being able to tell that much does not amount to knowing how it all works, and thus we are certainly in no position to provide electronic equivalents. Right now we have no idea how to make a computer with the world smarts of a dodo. Newspaper headlines about the wonderful things robots will be able to do in ten years or less are simply pipe dreams that cannot be backed up by any actual research in artificial intelligence.[6]

Whether this barrier can be overcome in principle I will not discuss here. For our purposes, the important consequence is that we have no reason to suppose that it will happen in the next decade or so. It is true that we have often achieved what pundits had declared impossible. The space program is one of our very best examples of that. But the history of our scientific civilization is also full of projects that we later discovered could not be realized. Among those projects we should include the proposals made in their youth by Tsiolkovsky, Goddard, and Oberth, the three pioneers of space flight, for a machine that could lift itself into orbit by its self-generated centrifugal force. As it turned out, the device violated the law of conservation of momentum.

In any event, the success of artificial intelligence is not likely to come soon enough to provide robots with what humans have to offer now to the progress of space science. Where it is inconvenient for humans to go, we must settle for what robots and teleoperators can do. But where we already know that humans can deliver the goods, it is not reasonable to snub them in favor of an uncertain technology. These remarks are not intended to argue against the development of more sophisticated teleoperators and robotics. On the contrary: there are many environments where humans cannot yet go, and will not go for a long time, and others where they should never go. If machines can go in our stead to those environments, we are so much further ahead.

Nevertheless, opponents of crewed flight take a look at the costs of the ISS ($150 billion as of 2010) and point out that a lot of space science could be done for that amount. For example, the NEAR mission to investigate Eros, an asteroid that comes as close to the Earth as fourteen million miles, had a price tag of about $211.5 million, which is pretty standard nowadays. Apart from its scientific value, this mission may someday allow us to figure out how to divert from our planet a similar asteroid. It seems incongruous that for the price of *one space station* we could fly instead *between 400 and 500 interplanetary missions*!

Let us consider some of the important missions, scientifically and otherwise, whose funding was affected by the diversion of monies into what some believe is a human-run sinkhole in the sky:

1. NASA's "system of environmental satellites was at risk of collapse"[7] because the agency shifted to the Shuttle and the ISS $600 million from the Earth sciences.
2. NASA, for similar budgetary reasons, had downsized the next generation of the National Polar-Orbiting Operational Environmental Satellite System. In particular, it had stripped out "instruments crucial to assessing global warming, such as those that measure incoming solar radiation and outgoing infrared radiation."[8]
3. For $100 million of fine-tuning the Large Synoptic Survey Telescope (LSST) we could identify 90% of asteroids between 100 and 1,000 meters in size. And since the LSST is Earth-bound and thus is limited (can spot the asteroids that come closest only at dusk or dawn, when the Sun's glare may obstruct our vision of them), for $500 million we could place in orbit around the Sun an infrared telescope that "could pick up essentially every threat to Earth."[9]
4. For $400 million, the proposed Don Quixote would have fired a 400-kilogram projectile into a small asteroid to affect its trajectory. This would have helped us figure out how to deflect asteroids bound to collide with our planet.

Although these examples were current as of 2008—many scientists still see in them an inkling of how far space science could go if not for our mania to send humans into space (incidentally and fortunately, the Double Asteroid Redirection Test [DART] was launched at the end of 2021 to accomplish the sort of goal previously assigned to Don Quixote).

When the ISS was first proposed, most space scientists feared, with good reason, that, on the whole, the ISS was going to take money away from many space science projects. I say with good reason because that is exactly what happened during the construction of the Space Shuttle. As the new vehicle could not be brought in under budget, the space sciences suffered a double jeopardy. First, their funds for many science projects were transferred to the Shuttle. And then many other experiments were not performed because they had been rescheduled to go on the Shuttle, but the Shuttle was not ready. The two Shuttle disasters made the situation a lot worse. It seemed reasonable for scientists to want to ensure that a commitment to the ISS would not be underwritten on the back of space science.

According to the bioengineer and NASA adviser Larry Young, "NASA always uses research as justification for its large crewed missions, but once they are under way the engineering, political, and fiscal factors take over and the science constituency is often cast aside."[10] Weinberg is far more bitter: of five missions proposed to challenge and expand Einstein's general theory of relativity, only one was likely to survive. "This is at the same time," he says, "that NASA's budget is increasing, with the increase being driven by what I see on the part of the president and the administrators of NASA as an infantile fixation on putting people into space, which has little or no scientific value."[11] For Weinberg, what has happened to the Beyond Einstein program reminds him of the time when the most grandiose particle-physics project, the Superconducting Super Collider, which was being built in Texas, was scrapped by Congress because funds were needed to build the ISS. It should not be surprising, then, that a great many space scientists have opposed the new proposals to send humans back to the Moon and on to Mars.

As for the disadvantages of teleoperators and robots, the opponents of human exploration point out, we can send dozens of dumb robots or clumsy teleoperated contraptions to take on the sundry jobs a human could theoretically do in space. Certainly, we will fail far more often, but the failures will not be as costly or devastating; we will save money, we will get the job done, and we will be forced to improve our science and technology. It is not just that we can try with machines again and again until we get it right, but also that we can divide the tasks humans would have performed into many simpler tasks and then try to accomplish those with swarms of new machines.

And let us not forget that machines have traveled tens of thousands of times further than humans have ever gone. What sense does it make to restrict exploration to dipping our toes when we could swim across the English Channel?

Still, the proponents of human exploration have a point—or rather, several. First, the significance of the science done in the ISS has improved greatly, as we saw in the case of space biology in Chapter 6. It has more recently carried out research also in astronomy, astrobiology, space weather, and fundamental physics.[12] The most notable ISS experiment has been the Alpha Magnetic Spectrometer (AMS), designed to help detect dark matter and tackle other crucial questions about the universe. It would not have done well in a satellite for it has large power and other technological needs. It has provided interesting results concerning the presence of dark matter, as well as large amounts of high-energy positrons in cosmic rays.[13]

The proponents would likely acknowledge the obvious point that the Shuttle set space science back many years. Even the head of NASA thought it was a mistake.[14] But they might point out that, in fairness, the Shuttle was a very poorly designed system, not a good example of how human exploration should be done. It was essentially a very large and heavy glider taken into low orbit by a partially reusable combination of solid and liquid fuel propulsion engines. It was at its worst when it ferried satellites into orbit, a task to which it was devoted through the 1980s, and for years afterward. Suppose that I am eating dinner in Tokyo next to a sumo wrestler. Across the dining room a friend asks me for the small pitcher of warm Saki. I could get up and take it to him. It would require me to spend a certain amount of energy, but not much, and to incur a certain amount of risk, but not much. Or I could hand the pitcher to the sumo wrestler and then carry him, Saki in hand, across the room so he can give the Saki to my friend. The energy required is far greater, as is the risk involved, both to the sumo wrestler and to me. Indeed, that is an extremely expensive and unsafe way to ferry satellites into orbit (or carry Saki across a dining room). To make matters worse, the Shuttle was an extremely complicated machine, with more ways to fail than anything else flown before or since.

Was there a better way to fly people into space? Of course, there was. Otherwise, we would have never landed a man on the Moon and returned him safely to Earth "before the decade was out" as President Kennedy challenged us to do in the early 1960s. And when we went to the Moon, let us recall, there were also many complaints about the poor quality of the science that was expected to result. Those complaints were very misguided. The Apollo astronauts made great contributions to science—contributions, for example, to our understanding of the origin and evolution of our planet.[15]

That Apollo era saw the flowering of machine exploration as well, a flowering that continued through the mid-1970s, until NASA put its rockets in mothballs and forced practically the entire U.S. space program into the coming "space bus"—the Shuttle. That era also gave us a reliable and relatively inexpensive space station, Skylab, which burnt in reentry when its orbit decayed, since NASA no longer had rockets to move it into a higher orbit. The Russians, with a somewhat more modest program also based on rockets, had a very active exploration of the solar system and a station of their own: Mir. Thus, the Shuttle hampered both machine and human exploration. Just before that fateful choice, NASA had built and successfully ground-tested a nuclear rocket engine (NERVA) that was to take us to Mars by 1981.[16] Just as we were ready for greater adventures and technological challenges, NASA made the "safe" choice instead.

It seems sensible, then, to explore with people and with machines, as long as we have a sensible means again for ferrying astronauts into space. The first practical step was to jettison the Shuttle (2011) and use Russian rockets instead, while trying to develop heavy-lift rockets of our own.[17] We are also encouraging private undertakings such as Spaceship1 and all the others mentioned in Chapter 2, which promise far cheaper access into space for astronauts.[18]

We should realize that human exploration is once again poised to help advance the cause of space science. I am referring to the search for life, or for fossils of life, in Mars. Rovers and other machines are already exploring the planet, finding evidence of water, laying the groundwork for when human geologists and paleontologist finally arrive, for many realize that their presence will be needed in this extraordinarily significant endeavor.

Space Exploration in the Long Run

How far can the human horizon in space expand? Space enthusiasts should not assume that this is merely an incremental matter. Let me consider first the possibility of colonizing the solar system, and then of expanding into the galaxy.

At Home in the Solar System

Colonizing the solar system requires that we solve the serious problem of prolonged exposure to radiation. Better shields, faster spaceships, and piles of dirt on top of our outposts should help a great deal. Better shields are within our technological means, as we saw in Chapter 6, and piling dirt on top of, say, inflatable dwellings should not present any major obstacles. Faster spaceships have been on the drawing board for a long time: fission rockets, fusion rockets, ion propulsion rockets, laser-propelled rockets, and even solar sails. Fission rockets and ion propulsion rockets would use tried-and-true technologies. Ion propulsion, for example, works on the principle of particle accelerators: you take a charged particle and you accelerate it by means of electromagnetic coils; the particle picks up a very large exhaust velocity and, by Newton's Third Law, it propels the rocket forward. The solar sail would use the solar wind to move around the inner solar system with great ease and at great speeds. It would consist of very large sails made of very thin sheets of metal that would move under the pressure of the solar wind. All these three systems could in principle reduce a trip to Mars from about seven months, on the average, to perhaps as little as three months or less.

We also need to solve the physiological problems created by microgravity. Perhaps the best solution is simply to create what is sometimes called "artificial gravity," which is not more than an application of the equivalence between acceleration and gravitational attraction. If a ship accelerates at a rate of approximately 10 m/sec^2, which is the value of 1 g, a passenger would feel as if on the surface of the Earth (a favorite illustration is of an elevator that ascends to the heavens unbeknownst to an unsuspecting passenger; as the elevator accelerates, its floor comes up to push against the passenger's feet, which leads to the feeling of being drawn toward the floor).

Another way to achieve the same result is to construct very large ships or habitats that rotate. At the appropriate rate of rotation, someone on the ground floor—next to the outer shell—would experience a centrifugal acceleration equivalent to the Earth's gravity. The reason this technique is not used on present spaceships is that, since they are relatively small, they would have to rotate extremely fast, which would subject the astronauts to Coriolis forces (if you are an astronaut, your head would feel a certain acceleration and your stomach another). Those effects, in addition to the extremely fast rotation, may play havoc with the control of the ship. Large structures on the scale of Gerard O'Neill's space colonies would not have such problems, but using a space colony as a spaceship might require seemingly prohibitive amounts of energy, at least in the near future (but see later). The most reasonable alternative, as we have seen before, would be to connect the spaceship to a cargo module by a very long tether, and to have the two rotate around their common center of gravity (the rocket engines). If the distance between the spaceship and the axis of rotation is 200 meters, for example, we would have the sort of acceleration that a gigantic spaceship or a small space colony might enjoy.

Human Expansion throughout the Galaxy

With present technology a trip to the nearest stars would take tens of thousands of years. Perhaps with an extension of our present capabilities we may be able to cut the journey to only a few centuries. Unless a truly fantastic technology for suspended animation is discovered, the trip would have to be completed by the descendants of the astronauts who begin it. Under those conditions the best way to travel to the stars might be to turn one of O'Neill's colonies into a vehicle and set it to depart from our solar system. But surely, some critics might say, people in their right minds would not wish to take their space colony into interstellar space for journeys that would last thousands of years—although how preposterous the idea is will have to be determined by a level of technology and an abundance of resources in interstellar space that we are in no position to predict now. At any rate, even people in their right minds may consider precisely such a journey if they knew of some unavoidable catastrophe that was to befall the solar system, or for other reasons that we may not fathom at this time.

Nevertheless, in journeys so long that only the descendants of the original travelers can complete them, a successful outcome may seem remote at best. Accordingly, some feel that the next great barrier to space exploration is the development of technology that would permit interstellar travel during a human lifetime. Since the closest stars are at least four light years away, and our galaxy is about one hundred thousand light years across, we would need starships that achieve velocities close to that of light.

Many scientists, however, believe that Einstein's special theory of relativity makes it unfeasible to accelerate spaceships close to that of light, whereas accelerating spaceships beyond the speed of light is simply forbidden by that theory. Nevertheless, as we will see later, Einstein's physics does not in principle preclude either of these options.

Let us consider the first option. Although Alpha Centauri is only four light years away, the majority of stars of interest in the galaxy are tens, hundreds, or thousands of light years away. It may seem then that even if we could achieve relativistic velocities, traveling to the stars may take as long as the astronauts' life spans, or longer. Fortunately, distance and time are relative to the inertial frame of reference in which they are measured (in an inertial frame of reference the velocity is uniform). In a ship that travels at great velocity with respect to us, time slows down and distances shorten, even though the astronauts themselves detect no abnormality. At velocities close to that of light, 300,000 km/s, distances are so short (or alternatively, the dilation of time is so large) that apparently unbelievably long journeys became feasible. According to calculations by Carl Sagan, we could go to many interesting stars and come back in a decade or two, ship time.[19] Six years of ship time would go by in a round trip to Alpha Centauri (eight years Earth time), twenty-two to the Pleiades (800 years Earth time), and to the Galaxy of Andromeda, which is over a million light years away, the round trip would only take about five decades!

In the meantime, nearly three million years would have gone by on the Earth, and so the return may offer the astronauts more of a shock than what they might find in Andromeda. Most of us would not want to go on such a journey, but I imagine that the project would suffer no dearth of volunteers. The main problem, however, would be the energy required. At a constant acceleration of 1 g our spaceship would reach 99% of the speed of light in one year. But in reaching it, we would need to spend, according to some calculations, energy equal to the entire consumption of the United States during a period of a million years! Enthusiasts like to point out that the first spaceship that went to the Moon spent an amount of energy tens of thousands of times larger than what many societies used only a century earlier. Bernard Oliver, who frowned on the idea of interstellar travel, thought that the requirements would be of this order of magnitude.[20] Even if such calculations are off by an order of magnitude or two, we are talking about staggering amounts of energy.

Some skeptical theorists have thought that the project is impossible, anyway, because as the velocity increases, so does the mass (also according to the special theory of relativity). But a larger mass requires larger energies to increase the velocity, which then increases the mass, and so on. This continuous increase in the mass of the spaceship eventually defeats the attempt to increase its velocity: We never reach a velocity close to that of light.

I do not believe that this objection works, however, for it does not take into account that from the point of view of the ship itself the mass has not increased. On the contrary, as most ships are conceived, it necessarily decreases as the engine burns fuel.

The skeptics' suggestion is, again, that as the traveler approaches the speed of light, the mass increases so that it takes more and more energy to keep accelerating at the same rate, making it really impossible to travel close to the speed of light, let alone reach it. Such reasoning leads physicists like Lee Smolin to conclude that "her mass increases as she approaches the speed of light. If her speed were to match that of light, her mass would become infinite. But one cannot accelerate an object that has infinite mass, and hence one cannot accelerate an object to the speed of light and beyond."[21] Similar remarks are made by Brian Greene[22] and even by Stephen Hawking.[23]

I think this line of reasoning is misleading in two ways. First, once again, as far as the special theory of relativity is concerned, the mass and the corresponding energy requirements increase only from the point of view of the observers left behind on Earth. But from the point of view of the star travelers, who are at rest with respect to the ship, the mass of the ship does not increase at all, and therefore accelerating the ship is not particularly more daunting than it was at lower velocities. If anything, it is easier because the longer the ship accelerates the more fuel it uses, and therefore the more mass it loses, as I pointed out earlier. At speeds close to that of light, its rest mass should be considerably less than at the beginning of its journey, as long as you have the standard means of propulsion (i.e., shooting something out the back, which for the crew may be the standard amount of exhaust, even though it may seem extraordinarily massive for Earth-bound observers). In practice, of course, the faster a starship travels, the greater the resistance from the interstellar medium, which could become significant depending on the ship's design and other factors. But this is a different type of concern altogether.

Second, the reason why the ship cannot match the speed of light has nothing to do with the mass becoming infinite. What physicists like Smolin, Greene, and Hawking have in mind is Einstein's equation:

$$m = m_0 / \sqrt{1 - v^2 / c^2}$$

where m is the mass of the ship from the point of view of the observer, m_0 is the rest mass, v is the velocity of the ship with respect to the observer, and c is the speed of light.

As v gets closer to c, the term v^2/c^2 approaches 1. This means that the denominator approaches 0, which makes m approach infinity.

But m can never reach infinity for the simple reason that, if the velocity of the ship reached that of light, the denominator would become 0 and the function would be undefined. The problem is not that an infinite mass is physically inconceivable, but that the mathematical expression makes no sense.

The main insight is flawed, in any event. Infinite mass has nothing to do with the relativistic speed limit. The reason why a ship cannot accelerate to the speed of light is that Einstein's velocity-addition formula (based in part on the postulate that the speed of light is a constant) will always yield final velocities less than c.

If I am traveling in a ship at 0.5 c, the speed of a ray of light with respect to me, whether it goes toward me or away from me, still is 300,000 km/s. If I fire a probe that travels at 0.5 c with respect to me, the speed of that ray of light would still be 300,000 km/s with respect to the probe.

The result is that, according to Einstein,[24] in the special theory of relativity, I cannot just add the velocities of my ship with respect to the ground (v_s) and of the probe with respect to me (v_p).

That addition must be divided by the term $1 + v_s.v_p/c^2$. When I add the velocities (my ship's plus the probe's) I do not get c, therefore, but only 0.8 c.

This corrected interpretation of the situation (from the point of view of the astronaut, and the appropriate equations from the special theory of relativity) still seems to forbid travel at or faster than the speed of light. It leaves open the question of building a spaceship that comes very close to the speed of light, though.

There have been other attempts to prove that near-light-speed travel is impossible, and there have been many refutations of such attempts as well. Of the presently available starship technologies (available in theory, that is) some form of controlled fusion may offer the best hope to achieve relativistic speeds (though just barely about one-tenth of the velocity of light). The ideal apparently would be a matter-antimatter engine, for it would convert all of the fuel's mass into energy as the particles and antiparticles annihilate each other. A serious problem is how to produce the necessary amounts of antimatter without spending extraordinary amounts of energy. And if you do produce it, you then have to worry about how to channel it so it goes out the nozzle only; otherwise it will radiate in all directions. And there is also the already familiar difficulty that if you include all the fuel you need to keep accelerating the starship, then you need to build a much larger starship, which then needs even more fuel, and thereby an even larger starship.

To get around these problems, we might employ starships that do not carry their fuel on board but take it from their environment. The first such "design" was for an interstellar fusion ramjet that would scoop hydrogen ions from space, the Bussard ramjet.[25] Bussard's interesting idea was marred by several difficulties, especially that it would require a scoop 160 kilometers in diameter and that it would use a proton-proton fusion reaction that may work only in temperatures as hot as the interior of stars.[26] A modified version, the Daniel P. Whitmire's catalytic nuclear ramjet,[27] apparently solves some of the main theoretical problems (it works by scooping up the hydrogen ions and running them through a catalytic nuclear reaction cycle, i.e., a nuclear reaction that repeats itself again and again, and that returns to the starting point of the reaction extremely fast so a new batch of protons can be used to propel the starship).

One possible sequence of reactions would be, for example, that of the catalytic cycle of carbon-nitrogen-oxygen (CNO), which occurs in the thermonuclear reactions of very hot stars:

$$^{12}C + {}^{1}H \rightarrow {}^{13}N + \gamma$$
$$^{13}N + {}^{1}H \rightarrow {}^{14}O + \gamma$$
$$^{14}O \rightarrow {}^{14}N + e^{+} + \nu$$
$$^{14}N + {}^{1}H \rightarrow {}^{15}O + \gamma$$
$$^{15}O \rightarrow {}^{15}N + e^{+} + \nu$$
$$^{15}N + {}^{1}H \rightarrow {}^{12}C + {}^{4}He$$

As we can see, the hydrogen ions (protons) react with the carbon isotope (^{12}C) to begin the cycle, which, after utilizing a total of four protons, ends again in ^{12}C plus a helium nucleus that is expelled out the nozzle, thus propelling the ship forward. The positrons (e^{+}) react with the electrons that remain from the ionization process and liberate additional energy in the form of gamma rays.[28] A compelling feature of the Whitmire ramjet is that it can generate as much as 10^{11} megawatts, about 10,000 times the energy produced in today's world.[29]

Whitmire and others[30] have worked on the possibility of either electromagnetic or electrostatic scoops of dimensions in the hundreds of meters, rather than kilometers, to reduce the immense proton and electron drag expected to affect a ship moving at relativistic speed (and that would be more efficient in the collection of protons). There are several practical problems with Whitmire's design, the most nagging of which is that the temperatures in his reactor might reach temperatures of *one billion degrees Kelvin!* This problem, however, is beginning to look less insurmountable. China's tokamak fusion nuclear reactor has reached temperatures of 180 million degrees for 10 seconds and of 120 million degrees for 101 seconds.[31] Its main operating temperature will reach 360 million degrees. The star-hot plasma is contained by a very strong magnetic field. This is not quite a billion degrees, but it is getting into the ballpark. Continuous improvements in the technology bode well.

There are also problems raised by the gravitational impact of a ship that moves through a medium with respect to which its mass increases extraordinarily (even if it does not change for the astronauts). It is possible that the ship may affect the structure of space-time in its path. There would be additional problems to describe mathematically the interactions between the ship and the environment, using the general theory of relativity, for the ship will exchange energy with the particles closest to it as it accelerates, which would then cause all sorts of difficulties for the calculation of the relevant masses. This problem may be lessened, however, by the action of the electro magnetic scoops that may funnel the overwhelming majority of particles in front of the ship, which then do not touch the hulk.

The feasibility of such potential technology assumes that the interplanetary space medium (ISM) indeed has a hydrogen density of one atom or ion of hydrogen per cc, as has been estimated for decades. The matter is not quite as straightforward, though. As Miles S. Hatfield (NASA's Goddard Space Flight Center) explains, "That one hydrogen/cc figure is a rough average; in fact the density is highly variable in different regions of the ISM, with some regions as dense as a million hydrogen atoms/cc, and other extremely rarified regions (like the Milky Way's halo) at more like .0001 particles/cc."[32] This may suggest that a Whitmire ramjet may have to travel in certain regions but not in others if it needs to accelerate. On the other hand, such ships may have specialized tanks to store hydrogen in rich regions to use later in places such as the halo of the Milky Way.

I cannot promise that these problems will be fully solved eventually, but even so Whitmire's idea seems very promising. I will return to it later in this chapter.

Faster-Than-Light Travel

The second fantastic proposal, the attempt to *reach and surpass* the speed of light, is even more interesting from the theoretical point of view. The basic intuition behind this proposal is as follows: the speed barrier applies only within the special theory of relativity, which requires the velocity-addition formula we have seen earlier, and which presupposes nonaccelerated frames, but it need not hold within the accelerated frames of the general theory of relativity, or within a theory of quantum gravity.

Indeed, the suggestions that have aroused the greatest interest in the last twenty years or so concerning travel faster than light both make use of the general theory instead. I will discuss briefly the most interesting two, not with the goal of determining which of these suggestions is more likely to take us to the stars in faster-than-light starships, but rather whether physical theory permits traveling faster than light.

The first suggestion is Kip S. Thorne's idea to use a Wheeler quantum wormhole to travel in a very short time to places that in normal space-time could be thousands or even millions of light years away.[33] Imagine that space-time is folded (e.g., in a fifth dimension in addition to the three of space and the one of time). That fold may bring close together, in that fifth dimension (or in so called "hyperspace"), regions of space that are extremely far from one another in the normal three space dimensions. It is as if we took a long cloth and brought close together the two ends. If the cloth were laid flat, the two ends might be separated by a distance of one meter, but now that we have folded the cloth, the two ends might be only, say, a millimeter apart. If we could only make a little tube that connected them across that millimeter, the trip from end to end would be far shorter. John Archibald Wheeler's proposed that, in extremely small regions (around a Planck length, 1×10^{-33} cm), strong gravitational quantum fluctuations create a sort of "quantum foam,"[34] in which we might find such a little tube, a "wormhole." The trick is to find one wormhole in the foam, to enlarge it so a ship can go through it, and then to keep it open so it will not crush the ship.

Let us consider all three aspects of Thorne's idea, which he developed upon a request from his friend Carl Sagan. The first problem is to find the wormhole. No one has ever detected one, and we do not even know if they exist. If none exist, or we cannot find them, an alternative would be to create one, as long as wormholes *could* exist. But can they? Wheeler's results came from his attempts to construct a theory of quantum gravity. Unfortunately, half a century later we do not yet have an adequate theory on the subject. It is difficult to say then, even if we take Wheeler's imaginative idea seriously, that physical theory does not forbid faster-than-light travel. The special theory of relativity certainly seems to forbid it. Does Wheeler's joining of the general theory and quantum theory somehow bring to light some limitations to the special theory? It might, if that junction were true, but that is precisely what we do not know.

Since we do not have an acceptable theory of quantum gravity, we are simply in a state of ignorance. From that ignorant perspective, travel faster than light may or may not be permitted by the deep laws of the universe that we do not yet know. The situation would be similar to asking in, say, 1855, whether it is possible in principle for a ship to travel at 300,000 km/s. Nothing would seem to forbid such a feat, but only because Einstein's velocity-addition formula was still fifty years away from appearing in print. Some may believe that the situation is actually worse, since we have no trustworthy theory to give us even seemingly reliable guidance—in 1855, we had Newton's.

Some reason for optimism comes from the Casimir effect, which demonstrates that the vacuum is indeed teeming with virtual particles coming in and out of existence, as we would expect in Wheeler's account. Casimir suggested in 1948 that if two metal plates were placed micrometers away from each other in a vacuum, and in the absence of an electromagnetic field, some virtual photons would not appear between them because of their long wavelengths. In that case, there would be a greater density of virtual photons outside the plates than between them. This difference in density would result in pressure being applied to the plates from the outside, and thus they would move toward each other. The confirmation of the Casimir effect in 1958 is strong evidence for the hypothesis that virtual particles are prevalent in the vacuum, and many now see it as confirmation also that there is some sort of space-time foam a la Wheeler. Notice, however, that the Casimir effect seems accounted for within quantum theory and is, at the very least, neutral about possible interactions between gravity and quantum effects.

Suppose, however, that we do find or create a Wheeler wormhole. Unlike the wormholes that might exist as a result of black hole singularities (in which the matter that disappears into the singularity "tunnels out" to another universe or another part of the universe, and in which the tunnel would close quickly and the extraordinary gravity would crush any would-be traveler), Wheeler wormholes would be extremely small, of Planck dimensions.

It would be necessary to make them longer, so as to connect one of the entrances with some desirable destination, and wider and stable, so we could send our astronauts through them. How could this be accomplished? The favorite answer: exotic matter. Now exotic matter is truly exotic. Presumably it would have negative mass, or at least exert negative energy (it would push the walls of the wormhole outward). And of course, we have no idea whether it exists. But Thorne and his coworkers think that something like the Casimir effect might produce negative energy inside the wormhole to keep it open.[35]

Think back, however, to the description of the Casimir effect given earlier. If we think of the vacuum as having zero energy, then we could think of the volume between the metal plates as having negative energy. It is a relative assignment of sign given the description we choose. But it does not seem sensible to say, for example, that the virtual particles within that volume have negative mass, or anything of the sort. Those virtual photons have no peculiar properties compared to the virtual photons outside of the plates. So it is not as if we could go looking for some exotic matter to spoon into the wormholes. But perhaps what Thorne is after is what he calls "exotic fluctuations," which would create negative energies, and which Hawking presumably showed existed at the event horizon of a black hole. Such exotic fluctuations would account for Hawking's radiation. Nevertheless, a less exotic description is that the black hole pulls a member of a virtual particle pair inside the horizon while letting the other member escape into the universe, thus creating a glow of energy around the event horizon. We may then say that this positive energy is compensated by negative energy being sucked into the black hole (again a relative assignment of sign).

In any event, to expand a wormhole's diameter along these lines, it would seem that a great deal of energy would have to be concentrated into the small region of the mouth of the wormhole so as to create a pronounced space-time curvature in that region. Whether this would really lead to the desired opportunity for faster-than-light travel would have to be determined by a good theory of the interaction of gravity and quantum phenomena—that is, if we only had one.

Of course, some of these theoretical ideas may turn out to be correct. Perhaps new experimental work that concentrates large energies into small regions could confirm the existence of space-time foam, wormholes, and exotic matter (or at least some way of bringing about something akin to the Casimir effect inside a wormhole). But until such a time we will not really be in a position to say that faster-than-light travel is possible.

Imagine, nevertheless, that we do find or create a Wheeler wormhole, expand it, and ensure its stability. We are still faced with a major conceptual difficulty: an outcome of travel through the wormhole is that an astronaut would also travel back in time. You may return before you take off! This gives rise to all sorts of puzzles about landing on your infant grandfather and killing him, which would make it then impossible for you to be born and thus to go on the trip in the first place. This absurd consequence would be a possibility in an established wormhole, if we do find one, since it is a possibility in general for faster-than-light travel, as has been known for a long time. An example I recall from my student days was that if you had a gun that shot tachyon bullets, one such bullet could ricochet off the wall and kill you before you pulled the trigger (tachyons are particles that always travel faster than light, and thus do not violate special relativity since they never accelerate to the velocity of light).

Thorne offers an interesting illustration of time travel in a manufactured wormhole. He imagines making a short wormhole with one mouth in his living room and the other in a starship sitting just outside on his lawn. His wife takes off in the starship traveling at close to the speed of light. Obviously, the two mouths of the wormhole have different times, once her trip begins, as measured in a framework outside of the wormhole, although inside the wormhole the times remain the same. Thorne's wife returns some hours later (her ship time), although years have gone by on Earth. She has two choices. She can meet Thorne on the lawn upon the landing of the ship and notice how much he has aged. Or she can crawl back using the wormhole to a time *before* she left on her space journey. Of course, she might then meet her own old self getting ready to go on her space journey, with the risk of bumping into each other, which would prevent her old self from starting that journey, and thus from bumping into herself on the way back!

These paradoxes make travel back in time conceptually absurd, which renders this type of faster-than-light travel also conceptually absurd. But could the paradoxes be resolved? One suggestion is that unknown laws of quantum physics (or quantum gravity, or who knows what) prevent anyone, or anything, to travel to the past and create impossibilities (it is impossible to go back and kill your grandfather if you were never born because you killed him when he was an infant). This suggestion is not only ad hoc but mere wishful thinking.[36]

Another is that the astronauts traveling through a wormhole would not create any inconsistent "time loops" because they would actually end up in an Everett alternative universe (according to Everett, each of two possible alternative quantum states is real, although each is real in a different historical line, or world, or parallel universe; since each line will soon face new alternatives, the result is an exponential growth of alternative universes).[37] So you would not really land on your grandfather; you would land on your grandfather's equivalent in a different historical line. It is difficult to distinguish this physics from science fiction, but even a less cynical appraisal of this suggestion should let us see that we are no longer talking about going to the past but rather to a different dimension or universe that is almost like your past and landing (and killing) someone similar to your infant grandfather. It is also obvious that you never arrive at your destination, but at a planet around a star that is similar to the one you were trying to reach, except for being located in another dimension or universe. By the same reasoning, your chances of returning home are very dicey. Wormholes appear to be problematic enough without combining them with Everett's interpretation of quantum mechanics.

A third suggestion is that there are indeed many worlds, more or less a la Everett, but that some are destroyed by inconsistent "time loops." Our world exists because no one has killed his grandfather, or the like, in any travel to the past. Time-travel consistency would function as a selection factor for possible worlds. The problem is, however, that when time travel alters the past, it destroys a history that could be tens, hundreds, thousands, millions, or billions of years long. We do not know from when the fatal time traveler is going to come. Moreover, he may arrive tomorrow at noon, or a million years ago. And we would no longer be. In fact, in the second scenario, we would no longer *have been*. This suggestion does not seem to work well as a solution to the paradox.

Thorne proposes an apparently more sensible approach. He discovered that some round-trip trajectories through a wormhole may be perfectly consistent: a billiard ball may return and hit itself a glancing blow that will still permit its earlier version to go on the trip.[38] Presumably, since this loop is causally consistent, we no longer face a paradox. Paul Davies seems to agree and provides an interesting variant: a rich man travels back in time, meets his (young) grandmother, and unwittingly gives her information about stock prices in her future. She invests her money using that information, which leads to immense wealth for her and for her grandson. Davies claims that "[N]o paradox ensues here."[39] Surprisingly he finds paradox in a case essentially alike. A professor travels to the future, finds a mathematical formula in a book, returns to his time, and gives the formula to a student, who then publishes it. That is the publication that the professor reads many years later. But neither the professor nor the student created the formula. Thus, information has come from nowhere, or rather just from the time travel. In the earlier case the grandmother could not have created her fortune (or the student written the paper) without the foreknowledge made possible by the relevant time traveler (grandson, professor). Because of the association between information and entropy, Davies thinks that this "free" information is "equivalent to heat flowing backward from cold to hot."[40]

It seems to me that the situation is even direr than that. In Thorne's example a sequence of events, a history, leads to a future event that in turn causes a destruction of that very history and its replacement by another. Something had happened, and now it has not. But if it has not, how could the inconsistent causal loop arise in the first place? Those who have no trouble accepting something like Everett's many-worlds view, or the even fancier notions of string theory, perhaps are not bothered by this new paradox all that much. But it must be pointed out, as it is generally accepted, that the empirical evidence for the first is scant and for the second nonexistent. In Davies's examples, there is not even an original history to create the conditions for the consistent causal loop: the man is already rich (without his grandmother having made the right investments?), the professor already finds the mathematical formula (that no one has really invented?). The loop just is. Causality is violated. Jorge Luis Borges would be pleased. But on the basis of such physics, we do not have enough to say that faster-than-light travel is possible.

One point of logic needs to be considered. Even if causally consistent time loops did not fall prey to these objections, it is difficult to see how such loops fix the conceptual absurdity of going to the past. The paradox is not that every time we go to the past, we kill our infant grandfather, and so on. The paradox is that *we could*, accidentally or otherwise.[41] Thorne's proposed solution is that there may be consistent loops. But the danger still exists that, for example, the rich man's son, years after his father's time trip, indeed, years after his father's death, finds the time machine, pushes the wrong button, and ends up landing on his infant great grandmother, crushing her to death. That would make his family's history, including the presumed consistent causal loop, become nonexistent! The paradox has not been resolved.

One might think that if it is possible for someone from farther in the future to annihilate that consistent loop, then that consistent loop was part of a longer but inconsistent loop; but in that case the longer loop itself would have been eliminated, and thus we have nothing to fear from time travel. Therefore, we have no paradox. This response might make some sense if we accept the metaphor of a frozen four-dimensional space-time (like a vine made up of time slices of the other three dimensions). In that metaphor time is already all laid out and some Cosmic Pruner has cut out all the inconsistent loops from the cosmic vine. This would be as ad hoc as it is convenient. But the universe we experience unfolds in time, and we need a rather long causal sequence of events to create the conditions under which someone can go back to his past to wipe it out. That sequence of events, however, would have no particular marks to distinguish it from the one humans find themselves in already. Since the building of an actual time machine is presumably way in the future, if ever, we would not know whether we are in a real world (because either we will not invent time machines, or if we do the time loops they bring about will all be consistent) or in a world that one unexpected day will no longer *have existed*.

No wonder, then, that to save physics from absurdity, Hawking conjectured that the unknown laws of quantum gravity provide chronology protection, that is, that the universe does not allow time machines. This protection, he said, will "keep the world safe for historians."[42]

To say that faster-than-light travel is possible, then, we need to identify some relevant theories that permit it, or else to empirical evidence that, even in the absence of theory, suggests that possibility (e.g., people knew that flight was possible because they saw birds and insects fly, long before they had any theories that explained the flight of birds and insects). One problem with the theories of quantum gravity I have mentioned is that they have not been accepted because the empirical support is not there. I suspect the reason their proponents openly pursue such wild imaginings is that relativity and quantum theory presumably granted physicists the license to make unintuitive claims. Hypotheses about Wheeler wormholes, branes, the multiverse, and the like, it seems, do not sound any stranger today than, say, the wave-particle duality of light and matter did almost a hundred years ago. But I think there is a difference. When Einstein accounted for the photoelectric effect by suggesting quanta of light, his explanation was generally rejected, even by those who used his calculations. Bohr, for example, pointed out that Einstein's account was contradicted by many experiments that showed clearly the wave nature of light.[43] Eventually, Compton's X-ray scattering experiments made Bohr accept the dual nature of light, and this led to his famous principle of complementarity. The moral of the story is that physicists were forced by the phenomena to propose and accept otherwise extremely unintuitive views. Their experiments were their warrant. Perhaps other, more sensible views would have done the job, but no one proposed a persuasive one. Moreover, as I have argued elsewhere, the unintuitive character of their views, at least in the case of the principle of complementarity, was due to the general acceptance of a mistaken epistemology.[44] Einstein's theory of relativity, although not similarly prompted by experimental results, was nevertheless soon an important tool in contrasting our ideas with the world.[45] None of this is the case with the highly speculative ideas so much in vogue today, and as we have seen, those ideas do not permit us to say whether we will ever be able to go faster than light.

To rule on that possibility, we need, in addition to the requirements of theory or empirical evidence mentioned earlier, a way of travel that does not imply going back to the past. A way out of this difficulty is to realize that the speed of light operates as a limit only within the special theory of relativity. But within the general theory we may find ways to travel faster than light without going back in time.

Miguel Alcubierre argued that if we were to build an engine that contracts space-time in front of the starship and expands space-time behind it, we could accelerate the starship to a velocity arbitrarily higher than that of light (see Figure 7.1).[46] Since the local space-time for the ship would be flat, the astronauts would not violate the relativistic speed limit at any one point in their journey, although, from the perspective of the Earth-bound observers, the ship might be traveling much faster than light. The situation is similar to that of a man riding a sea wave on a surfing board: relative to the beach, he is moving very fast, but relative to the wave, he is hardly moving at all. By thus warping space-time, the ship may make a return trip to Vega, which is twenty-five light years away, in, say, three or four Earth years, from the point of view of Earth-bound observers, instead of more than fifty, as would be the case under special-relativity considerations. Alcubierre's arrangement also has the ship move only into the future, as airplanes and slugs do, and so we do not have to worry about time paradoxes.

Of course, from this theoretical possibility to building a starship with a "warp" engine there is a long gap. What kind of technology could possibly contract space-time? Alcubierre has suggested exotic matter, but we have discussed that enough in this work not to pin our hope on it. For the purposes of this book, however, what matters is that the desired processes are possible,

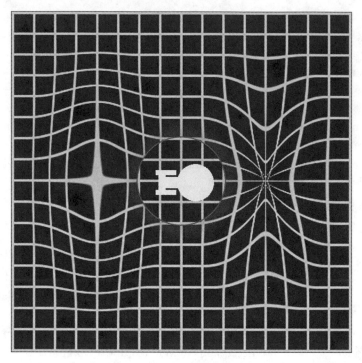

Figure 7.1. A starship has space-time expand behind it and contract in front of it. Eventually it may move faster than light, although it does not move relative to its local frame of reference. Its motion relative to other regions of the universe is the result of the properties of space-time (see color plate). (Illustration by Trekky0623. Public domain image)

given Einstein's theory of general relativity. Physics thus permit travel faster than light, as long as such travel does not include the possibility of traveling to the past. Alcubierre's proposal accords with the conceptual requirements, although, of course, there are no guarantees that such a spaceship will ever travel through interstellar space (see Figure 7.2).

I should mention, nonetheless, that Harold White and his team at NASA are trying to provide an experimental real-world proof-of-concept apparently by concentrating large amounts of energy into very small parts of space-time. This renewed interest in Alcubierre's suggestion has stemmed from White's calculations that the energy involved in a Warp Drive engine would be within the range of present interplanetary missions.[47]

Alcubierre's hypothesis may unexpectedly cast light on the question of dark energy. In 1918, Einstein wrote a note modifying his general theory in

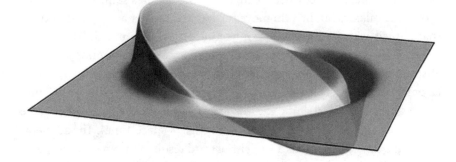

Figure 7.2. Regions of expanding and contracting space-time behind and in front of a central region. This combined action may move that central region (with a starship or a galaxy in its center) to speeds exceeding that of light (see color plate). (Illustration by AllenMcC. Public domain image)

that empty space would have gravitational negative mass.[48] He changed this notion of cosmological constant a year later. But perhaps the forgotten note deserves a second look. In today's most accepted model of the universe (the Lambda-CDM model), dark energy, which presumably explains why the universe's expansion is accelerating, is treated as the cosmological constant. As the universe expands, both regular and dark matter dilute, while radiation not only dilutes but redshifts. But the energy density of space itself remains constant through the expansion. That is, as there is more and more space, there is more dark energy to push galactic clusters away from each other. That is why the acceleration is increasing. When the universe was much younger, the density of matter and radiation was much greater, and thus the space between galactic and other structures was much, much less significant. Therefore, the cosmological constant did not play much of a role. But about five billion years ago, the universe had expanded and cooled enough that dark energy, as the cosmological constant, overcame the slowing effect of gravitation on the universe. Nonetheless, this is the picture most consistent with our observations concerning the expansion rate. In a few years, NASA's Roman Telescope (2027) and ESA's Euclid observatory (2022) will provide comprehensive observations from space to determine whether dark energy indeed fits the role of cosmological constant precisely.

All this is fine and good. Notice, however, that we could produce pretty much the same scenario with Alcubierre's account, putting one galaxy cluster in place of his starship. Let us consider the proverbial picture of the universe as an expanding bubble, where matter is in the bubble's skin. As the bubble grows, the galaxies on the skin move further and further away per unit time; that is, their velocity increases. Connect now this picture with Alcubierre's proposal for a starship. Of course, what is in accelerating motion is not an abstract universe, but its material components, local groups of galaxies, as embedded in their local frames of reference. In this variant, we picture space-time expanding behind the universe's bubble skin, while in front of the bubble skin it is "contracted" (literally the edge of space-time). The acceleration of the expansion is thus a straightforward result of the analysis of space-time in general relativity.

Alcubierre's approach seems consistent with a basic intuition of the cosmological constant approach, without any commitment to the quantum field calculations that lead to errors of great magnitude. That intuition is that vacuum energy (the energy density of the vacuum) could be thought of as the intrinsic energy of space and that general relativity predicts that this energy will have a gravitational effect. This energy, furthermore, will have negative pressure, which will cause the expansion of the universe to *accelerate* (i.e., the energy density puts pressure on the bubble skin of the universe to expand). But let us keep in mind that as the volume of the universe increases, the energy density of the vacuum per unit volume remains the same. This means that the total amount of energy inside the universe bubble increases, and thus the acceleration will likewise increase.

This extension of Alcubierre's thought allows us to explain the expansion of the universe without resort to enigmatic notions of dark energy, although some may miss the chance for the complicated mathematical fireworks and demanding conceptualizations expected from the latter.

In any event, it seems rather clear to me is that the physical expansion of humankind into the cosmos will vastly enhance our ability to preserve the dynamic character of science, while at the same time making it far more likely that a sun, some sun, will rise on the world of our descendants in a future so distant that most species on the surface of the Earth will have long disappeared. It is the process toward that long expansion that will make it possible to determine whether the relativistic velocities are technologically possible. I can only hope that we will set in motion the events that will ultimately allow our descendants to make that determination if they so choose.

I should close this chapter by offering my guess as to what approach would most likely allow for humankind to begin our interstellar migration in a question of decades, not centuries. There are two main requirements. One is the ability to achieve high relativistic speeds. The other is that we must first send a contingent of hundreds, perhaps thousands, of scientists who will be able to terraform some of the planets in the destination star systems. And, of course, we have to keep all those scientists alive and well during the trip, with artificial gravity and protection against radiation.

The answer (guess) lies in sending a space colony (a la Gerard O'Neill) propelled by several Whitmire ramjets (see Figure 7.3).[49] Our ship could be in the Alpha Centauri system in a bit less than five years—or wherever exoplanets, not too far away from Earth, seem promising. And once there, its scientific personnel will establish our interstellar beachhead. Human colonies could then follow in a reasonable amount of time.

This possibility of continuous expansion offers, then, a double bounty for our species. It increases our chances of survival, as we have seen. And it also preserves for a long time the opportunity to challenge our views of the universe. As Robert Goddard wrote in a letter to H. G. Wells, in 1932, "there can be no thought of finishing, for 'aiming at the stars,' both literally and figuratively is a problem to occupy generations, so that no matter how much progress one makes, there is always the thrill of just beginning."[50]

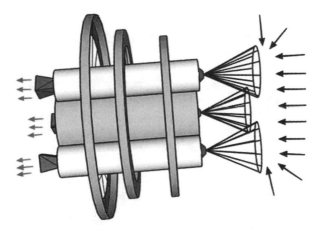

Figure 7.3. A space colony propelled by three Whitmire ramjets at relativistic speeds to begin human interstellar migration. Notice that the electromagnetic scoops in front of the ramjets create fields that take in all ions in the path of the ship (see color plate). (Illustration by Ruoyu Huang)

Notes

1. Kerr Than, www.space.com, September 8, 2007.
2. Reported by John Tierney, "Outer Space on Earth: NASA Should Try It," reprinted in *Detroit Free Press*, August 2, 2005, p. 7A.
3. The points presented in the first section of this chapter were initially published in Gonzalo Munévar, "Humankind in Outer Space," *The International Journal of Technology, Knowledge and Society* 4, no. 5 (2008): 17–25.
4. Paul Gilster, *Centauri Dreams* (Copernicus Books, 2002), p. 216.
5. John Hoegland, *Artificial Intelligence: The Very Idea* (MIT Press, 1985).
6. This point was made as early as 1972 by Hubert Dreyfus in his *What Computers Can't Do: A Critique of Artificial Reason* (Harper & Row). It was updated in a revised edition *What Computers Can't Do: The Limits of Artificial Intelligence* that includes a new introduction that reaches a similar conclusion but extends the 1972 analysis to include developments up to 1979.
7. This and most of the examples that follow are taken from George Musser, "5 Essential Things to Do in Space," *Scientific American*, October 2007, pp. 75. The present quote is from p. 70.
8. Musser, "5 Essential Things to Do in Space."
9. Musser, "5 Essential Things to Do in Space," p. 71.
10. *Science* 310 (November 25, 2005), p. 1245.
11. www.space.com.
12. "ISS Research Program," NASA. Archived from the original on February 13, 2009.
13. M. Aguilar et al. (AMS Collaboration), "First Result from the Alpha Magnetic Spectrometer on the International Space Station: Precision Measurement of the Positron Fraction in Primary Cosmic Rays of 0.5–350 GeV," *Physical Review Letters* 110, no. 14 (2013): 141102. doi:10.1103/PhysRevLett.110.141102.
14. Tracy Watson, "NASA Chief: Shuttles Were a Mistake," reprinted from *USA Today* in *Detroit Free Press*, September 28, 2005, p. 13A.
15. Cf. Chapter 4. See also S. G. Brush, "Harold Urey and the Origin of the Moon: The Interaction of Science and the Apollo Program," in the *Proceedings of the Twentieth Goddard Memorial Symposium* (1982), published by the American Astronautical Society; and "From Bump to Clump: Theories of the Origin of the Solar System 1900–1960," in *Space Science Comes of Age: Perspectives in the History of the Space Sciences*, ed. P. A. Hanle and V. D. Chamberlain (Smithsonian Institution Press, 1981), pp. 78–100.
16. http://ngm.nationalgeographic.com/2013/01/space-exploration/folger-text
17. The Constellation program, which NASA had proposed for returning to the Moon, was a high-tech update of the Apollo program, cancelled in 2010.
18. http://www.califcity.com/space_ship_1.html
19. C. Sagan, "Direct Contact Among Galactic Civilizations by Relativistic Spaceflight," *Planetary and Space Science* 11 (1963): 485–498.
20. B. M. Oliver, "Efficient Interstellar Rocketry," Paper IAA-87-606, presented at 38th I. A. F. Congress, Brighton, UK, October 10–17, 1987.

21. L. Smolin, *Three Roads to Quantum Gravity* (Basic Books, 2002), p. 79.
22. Brian Greene, *The Elegant Universe* (Vintage, 2003), p. 52.
23. Stephen Hawking, *The Universe in a Nutshell* (Bantam Books, 2001), p. 12.
24. Albert Einstein, "On the Electrodynamics of Moving Bodies," reprinted in *The Principle of Relativity*, with H. A. Lorenz, H. Minkowski, and H. Weyl (Dover, 1952) (republication of translation published by Methuen and Company, 1923). His equation as it appears in Section 5 of the article ("Composition of Velocities"), when the direction of motion of v and w is along the X axis, is $V = v + w/1 + vw/c^2$. See also Gonzalo Munévar, "Einstein y el límite de la velocidad de la luz," in *Einstein: Científico y filósofo*, ed. G. Guerrero (Programa Editorial Universidad del Valle, 2011), pp. 291–308.
25. R. W. Bussard, "Galactic Matter and Interstellar Spaceflight," *Astronautica Acta* 6 (1960): 179–194.
26. For an interesting discussion of this and other possible starships, please see E. Mallowe and G. Matloff, *The Starflight Handbook: A Pioneer's Guide to Interstellar Travel* (John Wiley and Sons, 1989), pp. 89–149.
27. D. P. Whitmire, "Relativistic Spaceflight and the Catalytic Nuclear Ramjet," *Acta Astronautica* 2 (1975): 497–509.
28. Other catalytic cycles such as ^{20}Ne would also be possible. The maximum energy generated by Whitmire's reactor would be around 10^{11} megawatts, about 10,000 times the amount of energy produced by the entire world today (Mallowe and Matloff, *The Starflight Handbook*, p. 114).
29. Mallowe and Matloff, *The Starflight Handbook*.
30. See again Mallove and Matloff, *The Starflight Handbook*, pp. 124–133.
31. GLOBALink, "'Chinese Artificial Sun' Sets New World Record."
32. Personal communication.
33. K. S. Thorne, *Black Holes and Time Warps: Einstein's Outrageous Legacy* (W.W. Norton and Company, 1994).
34. J. A. Wheeler, *Geometrodynamics* (Academic Press, 1962).
35. M. Morris, K. Thorne, and U. Yurtsever, "Wormholes, Time Machines, and the Weak Energy Condition," *Physical Review* 61, no. 13 (1988): 1446–1449.
36. Another approach is to nip all these speculations in the bud, by pointing out, as Jeffrey Barrett does, that "One might argue that there can be no threat of temporal paradoxes in GTR (General Theory of Relativity) since a particular mass-energy distribution and spacetime either is or is not a solution to the field equations—if it is, then the solution provides a model for all spacetime events" (personal communication).
37. David Deutsch, "Quantum Mechanics Near Closed Timelike Curves," *Physical Review D* 44 (1991): 3197–3217.
38. Thorne, *Black Holes and Time Warps*, pp. 508–516.
39. P. Davies, *How to Build a Time Machine* (Penguin Books, 2003), p. 96.
40. Davies, *How to Build a Time Machine*, pp. 102–105.
41. For some strange reason many physicists, including Thorne and Davies, thought at one time that the paradoxes of time travel arose out of the exercise of free will. Obviously that is not so.

42. As quoted in Thorne, *Black Holes and Time Warps*, p. 521. Vacuum fluctuations would destroy the Wheeler wormhole before it can become a time machine.

43. For a fascinating account see Stephen G. Brush, "How Ideas Became Knowledge: The Light-Quantum Hypothesis 1905–1935," *Historical Studies in the Physical and Biological Sciences* 37, no. 2 (2007): 205–246.

44. See, for example, "Bohr and Evolutionary Relativism," Ch. 3 of my *Evolution and the Naked Truth* (Ashgate, 1998).

45. Few ideas in the history of science have been corroborated as much as, say, Einstein's formula for relativistic mass (discussed earlier). Practically every time we use a particle accelerator, we confirm it with millions, perhaps billions of instances.

46. Miguel Alcubierre, "The Warp Drive: Hyper-Fast Travel within General Relativity," *Classical and Quantum Gravity* 11 (1994): L73–L77.

47. Harold White, "Warp Field Mechanics 101," presented at the 100 Year Starship conference in Atlanta, 2011.

48. Mentioned in Daniel Oberhaus, "A New Theory Unifies Dark Matter and Dark Energy as a 'Dark Fluid' with Negative Mass," *VICE*, December 6, 2018.

49. This guess is based on the possibility of solving the potential problems of Whitmire ramjets. A new addition to them is the notion that the proposal for the magnetic scoops might be too optimistic, as argued in an upcoming paper: Peter Schattschneider and A. A. Jackson, "The Fishback Ramjet Revisited," *Acta Astronautica* 191 (February 2022): 227–234. On the other hand, new developments may drastically reduce the need for magnetic coils, a possibility suggested by the 2022 Zap Energy's FuZE-Q reactor. "A Pinch of Fusion," *IEEE Spectrum* (January 2022).

50. Quoted in www.spacequotations.com/rocketryquotes.html

8

The Search for Extraterrestrial Intelligence

Are we alone in the universe? Is it really possible that no sentient being on a faraway planet ever contemplated the stars and felt awe? That only humans ever wondered about the nature of the universe, or pondered whether similar beings might be asking similar questions? In the view of some scientists, it is extremely parochial to suppose that we are alone—one more instance of the syndrome that once made us believe that the Earth was the center of the universe.[1] According to those scientists, we have no more reason now to believe that we must be the pinnacle of creation than we had once upon a time to believe that the Earth was so special.

Thus begins the reasoning that takes them to the conclusion that extraterrestrial intelligences (ETIs) are likely to exist, a presupposition without which the search for extraterrestrial intelligence (SETI) would make little sense. This does not mean, however, that the proponents of SETI advocate the building of starships at all. Indeed, many of its practitioners believe that star travel is not very likely, at least not for a very long time. They urge instead that we scan the skies for the radio or laser signals of other advanced species.

Success in their mission is seen by SETI proponents as of such extraordinary importance that at some point they proposed Project Cyclops (1971), a very elaborate, and expensive, array of radio telescopes to carry it out.[2] Their proposals were not received with much sympathy by those who control the purse strings, and thus over the years they had to content themselves with ever-meager levels of support (from tens of billions for the proposed Cyclops to less than two million per year in actual funding, and then to nothing). But what seemed like a deplorable situation to them appeared far too exorbitant to opponents of SETI. For in the view of such opponents, the very foundation of SETI, that ETI probably exists, was not only unwarranted but preposterous. U.S. Senator William Proxmire gave the program his Golden Fleece Award, for being the most inane waste of taxpayers' money. Eventually NASA cut SETI off its budget altogether. But the program lives on, bolstered by the privately funded SETI Institute and by the ingenuity and goodwill of many contributing scientists.

The Dimming of Starlight. Gonzalo Munévar, Oxford University Press. © Oxford University Press 2023.
DOI: 10.1093/oso/9780197689912.003.0008

Motivation for SETI

First, there is the incredibly large number of stars. This galaxy alone contains over one hundred billion, and there may be at least one hundred billion galaxies. We do not know how many of those stars have planetary systems, but most theories of star formation would encourage us to believe that planets are rather common. As we saw in Chapter 5, this optimism has been justified by the recent discovery of thousands of planets around other stars, with the data for thousands more being cataloged. Most are Jupiter-size planets, but the number of rocky ("terrestrial") planets has been increasing, although many are either much larger than Earth or too close to their parent stars to support liquid water and thus harbor life, presumably. The feeling is growing that the discovery of a "twin Earth" is imminent—indeed, perhaps as of this writing, a telescope is capturing the exciting evidence!

It is very convenient for SETI that the average stars may live longer than ten billion years. Since it has taken about four and half billion years to produce a technological civilization on this planet, it is encouraging to know that the stars that live long enough are also the ones most likely to have planets in the first place.

From here on, matters generally become far more speculative. Those who are in the business of making probability estimates for SETI often use the so-called Drake's equation (named after Frank Drake, the astronomer who first proposed it). According to this equation, the number of intelligent civilizations in this galaxy is equal to the product of the rate of star formation, the percentage of favorable stars, the number of planets around such stars, the fraction of Earth-like planets among those, the fraction of such planets in which life begins, the fraction of planets with life in which intelligence develops, and then the number of planets with intelligence in which technological civilizations arise. This product is then multiplied by the average longevity of a technological civilization.

We believe that in this galaxy the rate of star formation is about twenty per year. And the existence of other planets is now established, although the rate of planet formation has not been precisely established, but it appears to be high. As we progress through Drake's equation, however, the estimates are not as well grounded. This situation does not prevent SETI enthusiasts from assigning optimistic probabilities to every factor. One often hears, for example, that once life begins on a planet, intelligence is very likely to result eventually. Such optimism surely deserves examination.

Self-Reproducing Automata and the Impossibility of SETI

The fictionalization of space exploration, from H. G. Wells's *War of the Worlds* to *Star Trek* and beyond, has consistently offered us a vision of a future populated by alien civilizations.[3] The scientific underpinnings for that vision were most notably defended by Carl Sagan. That future vision is often supplemented by another: a future in which machines take their place alongside other sentient beings, as famously exemplified in all sorts of novels and films, including *Terminator*, with Austrian-actor-turned-California-Governor Arnold Schwarzenegger in the lead. The scientific underpinnings of that vision owe much to extrapolations of the work on computers by the Hungarian mathematician John von Neumann. It is not widely realized, however, that von Neumann's alleged proofs for the possibility of self-reproducing automata (SRAs) create a conflict between these two visions of the future; for the application of von Neumann's SRAs to space exploration seems to lend support to a famous objection against the existence of alien civilizations by the Italian physicist Enrico Fermi.

This chapter discusses the most critical philosophical and scientific assumptions involved in the proposal to explore space with von Neumann's SRAs. It argues that such proposal depends on von Neumann's mistaken metaphor of the genome as a computer program and is, thus, doomed to fail. It discusses also other suggestions to create exploring SRAs. Although they are ingenious, they are unlikely to succeed. Sagan's quest to search for advanced alien civilizations may thus remain to fire our imaginations one more day.

Sagan and Fermi

Ironically, the opposition to SETI is buttressed by the key assumption of the SETI proponents themselves: Carl Sagan's so-called Principle of Mediocrity.[4] The Principle of Mediocrity asserts that the Sun is a typical star in having a planet like the Earth in which life could arise, that terrestrial life is typical in having produced intelligence, and that human intelligence is typical in giving rise to a technological civilization.

Presumably, Copernicus taught us humility when he argued that the Earth was not privileged but average, and later astronomy reinforced the lesson by discovering that the Sun itself was merely an average star in an average galaxy. By extending the Copernican lesson, the reasoning goes, we should learn to be humble about our own position in the scheme of life. The Principle of Mediocrity thus purports to recognize that humanity and the conditions that have brought it are about average. In their arguments, the opponents of SETI stretch this principle slightly to add that a technological civilization, if any exists, should be typically expansionist. As a result, they are able to produce a variety of "impossibility proofs" against the existence of ETI.

If intelligent extraterrestrials do exist, Enrico Fermi once asked, why aren't they here? Some of today's opponents of SETI realize that it may be unfeasible, even for advanced civilizations, to travel, or to send "uncrewed" probes, to the (probably) billions of planets in the galaxy. Their impossibility proof depends instead on a technology they believe is inevitable: self-reproducing machines.

Objection

A contemporary version of Fermi's argument is that, if the SETI proponents are right, there should be technological civilizations far older, and presumably far more advanced than ours; for many "typical" stars in our galaxy are billions of years older than the Sun, and thus, in some of their planets, intelligence should have sprung long before it did on Earth. Now, just as we expanded from our beginnings in Africa, any civilization capable of space flight is bound to expand throughout the galaxy. Furthermore, this expansion would take place in a short amount of time: traveling at one-tenth the speed of light, which is within human reach now, it would take only one million years to cross the galaxy. Thus, the ETIs should be everywhere in the galaxy by now. But they clearly are not here; therefore, there are no ETIs in this galaxy.[5]

Exploring the Galaxy with Self-Reproducing Automata?

John von Neumann supposedly proved that human beings could design a machine capable of making copies of itself.[6] Indeed, NASA scientists have investigated the possibility of using such machines to explore the galaxy.[7] Since we already have a mathematical proof that self- replicating machines can exist, all we need is the talent, effort, and money to create the technology. Thus, more advanced civilizations would surely have discovered the equivalent of von Neumann's proof and would have developed the appropriate technology by now. Once these machines arrive at their destinations, they endeavor to make copies of themselves, which then move on to the nearest stars, and so on, setting in motion a geometric progression, until they overrun the entire galaxy. But we have no evidence of such machines here; therefore, no advanced civilizations exist.

Von Neumann offered not one but five proofs. The first one, however, is the main basis on which all these speculations rest. Von Neumann knew that, through evolution, organisms often produce others more complicated than themselves, and he wondered whether machines could ever do likewise. He then pondered the first step in that direction: was it possible to program a computer to make a copy of itself? He imagined a robot floating in a vat full of robot parts. The robot could be programmed to pick up a part and identify it. Then the robot, which had a blueprint of itself, would look for the connecting parts, and would then begin putting another robot together the way a child puts together an erector set. Once the second robot was assembled, the first would pass on to it the self-replicating program (or set of programs, rather). By breaking down the task of self-replication into small, manageable tasks, von Neumann thought, an automaton could surely copy itself (Figure 8.1). This result led him to remark that there were two kinds of automata: artificial automata, such as computers, and natural automata, such as people and cats.

Self-Reproduction of Automata in Space

If the European Space Agency were to turn one of von Neumann's presumed SRAs into a starship to explore the galaxy, the SRA would have to stop somewhere to make copies of itself. When an SRA arrives at another world, however, it is not going to land in a vat full of parts. It will have to build factories to build the parts from the raw materials that it will mine. But the factories are themselves made of parts, so it will have to build other factories to build the parts to build the factories. The specialists in the field call this the "closure problem." They do not fear an infinite regress, however, for we know that the closure problem can be solved. And we know that it can be solved because our technological civilization solves it: we do send rockets to other worlds.

Such an SRA, however, will be an extremely complex machine, both in its computer programs and in its physical realization. Indeed, to build an SRA, including the starship in which it travels, would demand many of the

Figure 8.1. John von Neumann's conceptual proof of the possibility of a self-reproducing automaton: a robot is programmed to assemble parts into a copy of itself; it then passes the program on to a new robot. (Illustration by Peter J. Yost from NASA Conference Publication 2255 [1982])

resources of a technological civilization. Whether *we* can write a program that complex, and whether *we* can assemble a machine that sophisticated is open to question. But let us assume for the sake of argument that we can.

Imagine that one of these extremely complex SRAs arrives at a planetary system. Now, we do not yet know what other planetary systems look like, but some theories and recent observations suggest that they would be collections of relatively small rocky planets and gas giants. If an SRA were to come into *our* solar system, Jupiter and Saturn would not be good places to land, even figuratively, because it is unlikely that the SRA can fashion the needed parts and factories out of the hydrogen, helium, and the trace gases that can be found in their atmospheres. The moons of the gas giants are not the best bet, for surely a machine that complex may be presumed to need a variety of materials, including metals, for the task of self-replication. Unfortunately, the low density of most of these moons (less than 2.0 g/cm³) suggests that, with a few exceptions, they would not be good places to search for the necessary raw materials.[8]

On rocky planets like Earth, however, an SRA can find practically everything it needs. But even on rocky planets the SRA's problems are far from over. Small differences between the planets in astrophysical terms may lead to significant differences in density and chemical composition of the atmosphere. These significant differences will in turn make it necessary to adopt different strategies for mining and manufacturing. For example, on Earth the best way to treat a particular ore may be to throw it into a pot of boiling water. In Mars, the water would evaporate before the ore is settled in. In Venus, the pot itself might melt.

This is by no means a small problem. No matter how similar planetary systems may be to one another, we should still expect at least some small astrophysical differences between their rocky planets. Thus, the possible combinations of factors that would affect mining and manufacturing may be practically infinite. Therefore, the already extremely complex SRA would need, in addition, some general-purpose programs so that it can begin the task of making the parts for its progeny. But no one knows how to write such a general-purpose program, and there are reasons for thinking that they cannot be written at all.[9] The biggest stumbling block to artificial intelligence has been precisely the inability to write programs that exhibit a flexible response to run-of-the-mill environments, let alone to the incredible variety demanded of SRAs. Nor is there any assurance that this problem can be overcome in the foreseeable future.[10]

Nevertheless, if rocky planets are too heterogeneous for the SRA's needs, we may instead settle for a homogeneous environment where they can find all the raw materials in question: the asteroid belt. Whether most planetary systems have asteroid belts is unlikely to be determined for quite some time, but if they do, an SRA would be able to move from asteroid to asteroid, picking up metal ores here, carbon compounds there, mining and processing them all in a rather stable environment. An alternative would be available if other planetary systems were to have the equivalent of our Kuiper Belt, which has bodies as large as Pluto, and whose distance from the system's star, about a light day away, would be convenient not only for replication but for the task of exploration itself.

After assuming all this, we are now able to deal with the fundamental problem of SRAs. As von Neumann himself pointed out, the more complex a computer program is, the more likely it is to have errors. But the SRAs would be far more complex than anything we have ever imagined programming and building. These errors, furthermore, involve principally the task of self-replication. It is not a mere question about bugs in the gigantic program, but about errors of execution in manufacturing and assembling the many components, such as an alloy that is not quite up to strength, or a gear tooth that is slightly short and with a bit of wear will no longer catch another as it must, and so on.

Neither quality control nor error-detecting programs will solve this problem, for it takes only a small percentage of error to bring the task of self-replication to a halt. The SRA is already saddled with a computer program so complex that it seems difficult to imagine that we can debug it completely. But now we must add to it a program to equip the machine with ways of checking the complete specifications for all parts, all fittings, and all functions. Nonetheless even this added complexity does not solve the problem. For a program that can foresee all the possible ways in which something can go wrong, including problems as minute as a piece of dust or a loose screw, begins to look like a general-purpose program—which is one reason why astronauts are so useful in space. In a machine that complex, engaged in the extraordinarily complex task of producing another SRA, things can go wrong in more ways than we can imagine. A program that must deal with so many unknown contingencies is, again, a program that can deal with an open-ended environment. And that is where SRAs come to grief, once again.

But isn't it the case that living beings, which are very complex in their own right, also make errors in the copying of the information used in replication? If so, a proponent of SRAs may ask why error brings machine replication to a halt when it does not do so in the replication of living beings. The answer is as follows. In living beings, "errors" caused by mutations or recombinations serve to provide genetic variation in a population. Although mutations often prove deadly, in some cases the genetic variation allows the individual, and eventually the population, to adapt to changes in the environment. For example, scientists develop drugs to identify the HIV virus by its molecular structure and then destroy it. But the HIV virus has a high rate of mutation, which means that the drug will kill most of the viruses that have the targeted structure but spare those few that have mutated their structure somewhat in the interim. In a short time, the surviving viruses will fully occupy the niche left vacant by the demise of the majority, that is, the host body. As the environment changes, then, some of the members of the population may take advantage of past "errors," and the population lives on. This is generally how error can be adaptive for living beings.[11]

In the case of the SRAs, however, we must remember that the asteroid environment was chosen precisely because it would not change from one system to another. In that unchanging environment there is not much advantage in error. In SRAs, the bulk of the errors that concern us are precisely in the reproductive part, and thus they are maladaptive. When all is said and done, it seems that an SRA technology is not even a gleam in a scientist's eye.

Nevertheless, many scientists think that we can point to examples of self-replicating machines: trees, cats, and humans. These scientists already assume von Neumann's conclusion, that there are two kinds of automata—natural and artificial. They find that assumption reasonable because they believe that the genetic code is the equivalent of a computer program, and thus they conclude that living things are just the realizations, or executions, of their programs. This view is no longer popular among biologists, but we still need to see why it fails to help the case of the SRAs.

If we are to use analogies, however, we should stress the following: In the case of the SRAs, the machine must make a copy of itself and then pass on the program. Even if it passes on the program to a unit that is not yet completed, making the copy and passing on the program are separate, largely independent tasks. In the case of living beings, however, it seems more proper to say that they pass on the "program" first, and then, as result of that, the copy is made. This is not a small difference, for in living beings relatively simple accomplishments, for example a fertilized egg, can produce very complex organisms—the egg is chemically extremely complex, but simple relative to, say, the human child that will result from it.

Genomes and Computer Programs

An even more important point demands consideration. To picture the genetic code as a computer program is just to engage in metaphor, and this metaphor is highly misleading. The "instructions" of the DNA produce the expected results, for example specific proteins, cells, tissues, organs, and behavior, only because at every one of those levels such "instructions" can be expressed in appropriate environments; indeed, it is often the appropriate environment that will trigger the next stage in embryonic development. In a human, for example, at a certain time in the life of the embryo the normal development requires a certain concentration of sodium, and later, after the human is born, the attention paid to it, even being touched, is not only necessary for its survival but provides the stimulation needed for the central nervous system to grow properly.

In other words, the "instructions" of the DNA do not have meaning by themselves. This issue is similar to that of the meaning of words in the philosophy of language. It used to be thought that words had intrinsic meaning, but it is now generally accepted that the meaning of a word depends just as much on the context in which it is uttered: the meaning is given by the interaction of word and context, where the context may include a large variety of factors, including the relationship of the word to other elements of the sentence, the manner of its utterance, the social conditions that the speaker and the listeners take themselves to be in, and so forth. In embryonic development, the "program" makes sense—has meaning—only in that the maturing organism interacts with a sequence of appropriate environments. Those environments provide the biological contexts in which the "instructions" of the genetic code are instructions at all.

Now, that sequence is itself the result of natural history, that is, of a long series of interactions between the ancestors of that organism and the environments in which they evolved. There is a clear sense, then, in which a living being comes into a world that is already made for it. The world of the SRA, on the other hand, must be largely described in its program from the beginning. The meaning of the program must be made explicit beforehand. The development of a nervous system offers a clear contrast. In dissecting an animal, we may find that its nerve cells always exhibit a certain pattern and may thus imagine that pattern is contained in a blueprint in the DNA. Nevertheless, as the nerve cells grow through, say, a muscle tissue, they do not need to be guided by any such blueprint. They may simply have "instructions" to grow in the general direction of a chemical marker, until they make contact with a membrane that turns the "instructions" off. But the developing muscle cells will then constrain the manner in which the nerve cells grow, which will then have to grow around the muscle cells. The final pattern is the result of such contingencies, and there is no need for any blueprint whatsoever. Otherwise, we would face an uphill battle trying to explain how as complex an organism as a human being, an organic whole whose individual organs have billions of cells, can simply be the "hardware" construction of the instructions contained in the fewer than 30,000 genes found in the human genome.[12]

As the German molecular biologist Gunther Stent (1924–2008) has pointed out, a true genomic "program" would have a structure that "is isomorphic with, i.e., can be brought into one-to-one correspondence with, the phenomenon."[13] But one "of the very few regular phenomena independent of human activity that can be said to have a programmatic component is the formation of proteins." In this case a stretch of DNA will be "isomorphic with the sequence that unfolds at the ribosomal assembly site."[14] This programmatic element, however, does not go very far, given that: . . . the subsequent folding of the completed polypeptide chain into its specific tertiary structure lacks programmatic character, since the three-dimensional conformation of the molecule is the automatic consequence of its *contextual situation* and has no isomorphic correspondent in the DNA"[15] (emphasis added).

Stent thus explains how the same genes may produce proteins with different tertiary structures because of contextual factors that are independent of those genes. But different tertiary structures may exhibit different chemical properties that are crucial to further development, and so do different quaternary structures (Figure 8.2).[16] In an organism this contextual situation is nested within another contextual situation, which in turn is nested within another contextual situation, and so on.

To see how far this nesting of context goes, consider the following examples. One might think that genes determine the degree of "maleness" or "femaleness" in the behavior of animals toward those of the opposite sex, including not just sexual preference but also how aggressively the animal seeks copulation. And in some sense the genes do cause such behavior, but only within a context of development that takes place after gestation and even

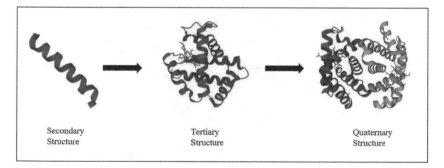

Secondary Structure

Tertiary Structure

Quaternary Structure

Figure 8.2. The same polypeptide chain (folded as a secondary structure) may fold on itself in a variety of ways to form different proteins, with different chemical properties due to the new and different interactions allowed between amino acids (as they come close together in the tertiary structure). The folding that leads to the tertiary structures depends on the biochemical context. An additional level of complexity, similarly created, is found in quaternary structures. (Illustration by Phillip McMurray)

after birth. In the case of rats, the litter in the uterus is made up of males and females. Female rats that lie next to males may try to mount other females later in life.[17] Moreover, when rats are born, the mother will lick the anogenital areas of her pups, but will lick the males longer, guided by their scent. If her olfactory sense is damaged, she will lick all equally, with the result that the males will be less likely to mount females as adults.[18] Rhesus monkeys raised in isolation from peers their age also fail to mount females as adults. The reason is that in play-acting with other young male monkeys they practice, among other things, the double foot-clasp mount characteristic of their species.[19] Without this social context, the genes for such basic biological function, reproduction, will not be expressed (Figure 8.3).

These considerations show that von Neumann's idea of the genome as a computer program is at best a very loose metaphor. Trees, cats, and humans are not natural automata. Like other living beings, they are instead complex creatures whose development depends heavily on context. SRAs, on the

Figure 8.3. Genes relevant to developing sexual behavior do not become "instructions" unless expressed in appropriate contexts, some of which are social. Young male monkeys act mounting behavior with others their age. If reared without male playmates, they will fail to mount females later in life. Testosterone does not mediate this social process. (Illustration by Nicole Ankeny. Reproduced courtesy of Palgrave Macmillan, London)

other hand, are determined by the programmatic character of their "blue-print," and their "mutations" do not fit the pattern that allows some living begins to have the "approval" of natural selection bestowed upon them.

Robots to the Rescue

Nonetheless, some space theorists have pushed ahead with plans for interstellar missions using new approaches to SRAs. One researcher, Robert Freitas,[20] has proposed sending to Alpha Centauri a machine about 200 times the mass of the already gigantic Daedelous starship, the British Interplanetary Society's concept of an advanced fusion-propelled ship, which captured the imagination of space enthusiasts in the 1970s and 1980s.[21] Freitas's ship would be populated by specialized robots, such as miners and metallurgists, with the ability to divide the tasks of replication into different categories. Freitas would try out this divide-and-conquer strategy in a Jovian-like system.[22] He believes a large-scale mining of a planet-size moon would provide the materials the machine would need to build a factory to replicate itself, at the rate of one replica per 500 years. The ore-processing problems would be overcome by the brute force of a machine that breaks up chemical compounds into its component isotopes. Once completed, the new SRAs could be fueled by scooping ^{3}helium from the giant planet's atmosphere.

It seems, however, that, if anything, Freitas's proposal shows the extraordinary complexity of the task. On Earth, even far more modest undertakings require decisions by humans, for things are bound to go wrong in unexpected ways. When the complexity of the task increases so dramatically, and when the environment will not be quite like the one envisioned before the ship departs from our solar system, things are even more likely to go wrong.

In response, Freitas has suggested principally semi-intelligent troubleshooting robots to replace those humans. As for the first trait, however, we do not know that we will be ever able to produce intelligent or semi-intelligent robots (What would the latter be anyway? Robots with a low IQ?). Even if the strong arguments against this possibility, which are readily available in the literature,[23] were mistaken and we could not rule out true artificial intelligence in principle, we should not conclude therefore that it is feasible. A woman cannot rule out in principle that she will wake up tomorrow and find herself thirty years younger—perhaps the combined action of 200 key genes caused by an improbable epigenetic event would bring about spontaneously what scientific medicine *might* accomplish in 500 years—but it would be unreasonable for her to expect it. Freitas is just waving his hands.

Computer enthusiasts like to point out that computers have become very "smart." Computers certainly compute a lot faster and do more things than they used to. But those abilities have little to do with true intelligence. Brains, real brains, that is, do not operate the same way as our admirable desktops and laptops. To arrive at a result, a serial computer may make millions of calculations, one after the other, in a second or so. But our brain must work in a different way, for the speed of transmission in neurons is of the order of 200 meters per second. This means that the decisions we make, often faster than serial computers, cannot typically proceed by serial calculation. We could try building robots that work in ways more akin to the parallel "processing" that takes place in animal brains. Neural networks along such lines are already able to imitate and sometimes surpass some abilities of human beings in restricted situations, limited to picking up certain aspects of grammar, for example. But even here we encounter two important restrictions.

The first is that a neural network is trained by feeding back to the network the amount of error between its output and a desirable output, say, distinguishing male from female faces. We give it rather large number of pictures of faces from both sexes, and it has the task of dividing them into male and female faces. By coming up with new ways of "weighing" features, given the feedback of its error, it improves, and eventually it can tackle the training set of pictures with a high degree of accuracy. It can also achieve a decent record of discriminating between male and female faces in new pictures.[24] But humans provide the training set; humans, that is, use their knowledge to train the computer to do the correct discrimination. In circumstances that differ from human experience, we can provide much less guidance. And when we send such an extraordinarily complex machine, or set of machines, to deal with new environments, the things that will go wrong will often include what we could not have anticipated. This nearly always happens when we explore.

When the German physicist Wilhelm Röntgen (1845–1923) entered his darkened cathode-ray lab in the late 1800s, he noticed a strange glow on the wall. His experience told him that such glow should not be there. It had to be energy of some form. But the equations describing the operation of his machine balanced: there should be no extra energy and thus no glow on the wall. The sense that there was something wrong led him to a series

of investigations that culminated with his discovery of a new form of electromagnetic radiation, X-rays, and required that a lot of work be redone.[25]
Although it would be unfair to demand that robots be as intelligent as a top
human scientist, the example nonetheless illustrates the ways in which real
brains detect and solve problems.

In short, we can train neural networks to make discriminations when we
can give them examples of correct and incorrect responses. This would be
increasingly difficult to do when we cannot anticipate the sorts of problems
that may occur, and when the complexity of those problems may be considerable. And in a new planetary system, trying to build a fantastically sophisticated and complicated machine will likely give rise to many new, complex,
and scientific challenging difficulties.

Moreover, a characteristic of intelligence, which Röntgen's example
illustrates, is the ability to tell relevant from irrelevant, significant from insignificant. This is the second important restriction. Animals achieve this
ability, in part, because natural selection has given them some basic emotions
that sound an alarm in their brains or at least tip the balance in pattern rec-
ognition, when matters of biological importance come across their gaze, or
hearing, or taste, and so on. As Antonio Damasio points out, human beings
with lower frontal lobe damage—an area of the brain where presumably
emotions inform reasoning—may continue to apply the "rules" of reasoning
properly but are no longer able to make reasonable decisions in aspects of life
in which they had been quite competent before their injuries. It is as if the
gears in their brains spun aimlessly now.[26] Perhaps someday an equivalent
connection to that between emotion and reason can be infused into robots.
But right now, we do not know that it can, and thus we should not suppose
that future research will make good our present hand waving.

Perhaps Freitas could receive help from self-repairing programs based on
genetic algorithms. For example, the "factory" could produce new robots
depending on the nature of the problem. If the standard-issue robots fail to
deal with a problem, a super maintenance program could take the computer
code for building those robots and introduce, in a virtual environment that
represents the archetype of the replica, say ten copies of that code with a few
"mutations." The resulting virtual robots would be made to "perform" in the
virtual environment. When one performs better than its rivals—for example,
it solves a little more of the problem than the others without causing additional damage—then its mutation is preserved into the next generation: ten

new virtual robots with new mutations introduced into them. The idea is that eventually a generation of virtual robots will fit the virtual environment. The "factory" would then produce actual robots that meet those specifications and send them to fix the problem.

Using genetic algorithms may indeed be a clever way to increase the autonomy of uncrewed starships at distances too great for human intervention. But such use would have serious limitations in the case of SRAs. It is one thing to have virtual environments that challenge the competing virtual robots to, say, climb a terrain with simulated large rocks and ravines. Another is to deal with a situation that may require scientific daring, a "gestalt switch," or that a robot by itself could not fix without the prior invention of a tool not yet a gleam in any engineer's eye. Of course, even humans may also fail in similar circumstances, when trying to make a replica of a starship under unforeseen circumstances. But the point is that such circumstances are likely to occur when dealing with such an extraordinarily complex machine in unusual circumstances. What are a robot's chances then?

Besides, genetic algorithms may not lead to the perfect match. They are restricted both by the virtual worlds they can produce and by the kinds of code available to them. If you took a gecko into Röntgen's lab and modified it by mutations, there is little guarantee that, even after millions of years, one of its descendants would be able to discover X-rays. Evolution in the direction of higher intelligence does not make sense for all sorts of organisms. It makes sense only for certain types of organisms who can afford a certain type of change in their metabolism.[27] Evolution, real evolution, does not guarantee a perfect match. It does not even guarantee an approximate match. It can only offer chances for improvement when circumstances vary—sometimes. After all, most species that ever lived are now extinct.

Furthermore, evolution occurs not only because organisms change to fit the environment. Organisms often change their environment, and that change may introduce new factors of selection. The case of actual, as opposed to virtual, changes in the starship may indeed introduce organic changes as well. That is, the conditions of interaction between the starship SRA and its environment may well change in unforeseen ways as well. But we could not then anticipate, and thus program, the changes the computer needs to make in the virtual environment that would be testing the virtual robots. As Pablo Picasso once said, "Computers are useless. They can only give you answers."[28] Of course, the prospect of such organic changes defeats, once again, the choice of a homogenous environment.

Nano-Robots

A new favorite solution is to appeal to nano-robots (sometimes called nanobots). Nano-robots, it is said, will cure cancer and all other sorts of diseases, will create new technologies, make our spaceships a lot smaller, and, best of all, replicate themselves. We should thus send a very small starship full of nano-robots that will make copies of themselves and of the starship. Indeed, the nano-robots would have to reproduce in extraordinary numbers if they are to build any machine large enough to be seen with the naked eye. The idea is to create a so-called assembler that would pick up an individual atom at a time and put it in the right place to make the right kind of molecule. Once assembled, the nano-machines would execute their programs and together produce, say, a starship.

This idea brings up von Neumann's proof again. The little robot picks up the parts (different kinds of atoms) and puts them together in the style of an erector set. Of course, the desirable atoms are not going to be floating in the nano-equivalent of a vat full of parts. They will be parts of molecules or be found in solution, and so on. We will have to "mine" them. This would not be easy for a nano-robot, for it would take a considerable amount of energy and finesse. It would require sensors, a way to store and apply energy, and so on. The number of atoms involved to carry out all those functions would make it a very clumsy tool. A good tool would somehow break the bonds the desired atom has created with those other atoms around it and then deposit it into its proper place in the molecule being built by the nano-machine. Perhaps we could solve the problem by some *macro-machine* that would blast the ores into individual elements. But the problem would persist at the point of assembly. I once knew a man who had a watchband accidentally tattooed on his wrist. He was an electronics technician with a penchant for disregarding safety rules. On one occasion he stuck his hand into a machine with a powerful electromagnet, not bothering to take off his metal-band watch. The powerful magnetic field slammed his wrist against that part of the machine and the current marked him for life. Likewise, the nano-robot is going to interact with the assembly being built, which will not sit there passively like the robot under construction in von Neumann's scenario. Another problem is that the atom being moved will form a bond with the nanobot's "finger"; that is, it will stick to it.

In the opinion of Richard Smalley, awarded the Nobel Prize in Chemistry in 1996, these problems are insurmountable for nano-robots.[29] Many other distinguished scientists share the same opinion.[30] And, in fact, no one knows how to make the famous "assembler" proposed by Eric Drexler in 1986.[31] Indeed, the problem of stickiness already plagues micro-electromechanical systems; it is bound to get much worse at the nanometer level. As if that were not bad enough, the star-trekking nano-robots will be operating in the vacuum of space, where the Casimir effect is likely to play havoc with their operation. The Casimir effect comes about when two thin plates are brought together at very small distances in the vacuum. As we saw in Chapter 7, the vacuum is not exactly empty but full of virtual particles coming in and out of existence. The narrow space between the plates rules out virtual electrons of certain wavelengths, and this creates an imbalance in the density of virtual electrons within and outside of the plates, which in turn pushes those plates closer together. Nano-robots in space would be plagued by "stickiness."

It is not very clear how one would "program" a nano-robot to perform its tasks: it must not only make a copy of itself but also carry out, in synchronization with many other nano-robots, the construction of much larger, perhaps macroscopic structures. The difficulty does not find parallel in the world of living beings, for they do not reproduce themselves by making their own replicas one atom at a time, but rather by serving as templates for the assembling of similar polymers that fold three-dimensionally in useful ways. The action of these polymers then leads to the creation of cells, which in turn form parts of organisms. And we have seen earlier that such genomes do not really function like computer programs. Thus, using nano-robots does not avoid the previous problem, but instead adds complications to it.

Insofar as the nano-robot would have a program to accomplish what living beings do without it, such a program would have to go well beyond little tricks like having a molecule "remember" a shape, that is, revert to a previous alternative configuration under certain conditions. Assembling a replica and helping in the construction of a starship are complicated tasks for something made up of very few atoms, relatively speaking. Nor is it clear how exactly the nano-robot would absorb and transform energy to carry out its tasks. It seems that machines of a larger scale, possibly macro-machines, would have to hold the relevant programs and infuse the necessary energy into their herds of nano-robots. We already saw that we might need a macro-machine to "pulverize" ores into individual elements. Moreover, part of the function of an exploring starship is to send information back to its home planet; but electromagnetic transmission at such distance may well require very large structures. All this in turn makes it likely that the means of propulsion will also involve large structures. These and other considerations begin to rule out the possibility of building starships the size of a Coke can, let alone smaller.

Of course, today we already find ways of programming computers without giving them an exhaustive account of every possible action they might take. For example, we can buy rather cheaply small, round floor-cleaning robots. They do not have the floor plan programmed into them, including, say, the position of the legs of a chair. If the robot encounters an obstacle, it executes a simple maneuver to get around that obstacle. This new approach might be coupled with the ability to cooperate with other robots the way social insects do to create hives, defend them, and so on. The robots in question would cooperate instead to make the replica of a starship.

Although it is true that every long journey begins with one step, not every step leads to a long journey. Learning to swim a lap at the local pool does not warrant optimism about swimming across the English Channel, let alone across the ocean. That computers can do a few rather simple things should not let us conclude that they can achieve the extraordinary level of sophistication needed to make a working replica of a starship. Maybe they can, but at this time this is more a dream than the assurance of a practically inevitable technology that any advanced technological civilization is likely to have mastered. Besides, let us remember that insect hives, like complex organisms, result from the unfolding of biological processes within larger biological contexts that are themselves the product of natural history.

Much of the optimism for these sorts of technological proposals comes from computer simulations. One problem with simulations, apart from the quip that you cannot eat the simulation of a good meal, is that they deal with theoretical scenarios, in the sense of "theoretical" that connotes abstraction from details, which are often unknown. We just do not see how "in principle" this or that could *not* be done eventually. But God, as they say, is in the details. In the 1970s, for example, great hopes for the cure of cancer were raised by the development of recombinant DNA. And many specific potential cures were indeed proposed soon after. Theoretically they should have worked. But they did not. We have learned in the intervening years that the behavior of cancer cells is much more complex than we could have then imagined, for a tumor has many clever ways of protecting itself.

We should also keep in mind that, when projecting from small successes to extraordinary feats, problems of scale could easily arise. If we want to make a dog the size of an elephant, we will have to make it look a lot like an elephant, with thick bones, and so on.

And even if we perchance figure out how to make new molecules that self-replicate, we do not know that those molecules would work well as nano-robots in a starship. Living cells replicate. DNA replicates. But neither seems a promising candidate to solve the problems that may arise in the building of a starship.[32] Solving those problems would require experience and ingenuity that we cannot program into large computers. What guarantee do we then have that the difficulty disappears just because the computers are made extremely small? Those nano-hands seem to be waving extremely fast.[33]

Assumptions

Several of the assumptions in this section may be challenged. For example, the starship SRA need not be as complex as envisioned so far: a machine with millions of parts. A new manufacturing technique called "rapid prototyping" can build rather complicated three-dimensional gadgets in one piece. It works by laying down a substance into the proper shape. When the substance congeals, we have a one-piece stand or handle, and so on.[34] If this could be done with the right kinds of alloys, we might be able to reduce substantially the number of components involved, even if we cannot build the entire hull of the starship in a single block. We may also improve quality control and preserve the structural integrity of the ship by means of electromagnetic identification (ID) tags properly placed to give us a three-dimensional image of the ship, a technology descended from the radio-frequency ID (RFID) tags of today, which either transmit or respond to radio signals, thus providing information about the location of objects or materials.

Self-healing starships would simplify matters even more. The self-healing is made possible by myriad microcapsules full of a special sealing material. Any crack in the hull, for example, would cut across several of the microcapsules. The sealing material would then be released and fill the crack. These examples give probably just a peek at the extraordinarily clever technologies likely to come the way of spaceship builders in the decades to come. They will surely reduce somewhat the complexity of starships.[35]

Reducing complexity, however, is not the same as eliminating it. The communication, navigation, and propulsion systems cannot be of one piece with the hull under any technology so far envisioned. Nor can a similar approach to their construction be undertaken, since each of those systems is made of parts that perform very different functions, undergo different stresses, and withstand sometimes drastically different temperatures. Moreover, the three-dimensional printing machine that makes the hull will itself have to be carried and then replicated. This adds complexity. The "meta machine" that builds it will then have to be replicated as well! And so on.

Wisdom Concerning Self-Reproducing Automata

A technology of self-reproducing machines may or may not be feasible, but we do not have sufficient reason for asserting that it is, and even less to conclude that it is practically inevitable. But an impossibility proof requires at least practical inevitability. Therefore, the impossibility proof fails.

Even apart from this problem, this impossibility proof would suffer from other serious defects. Unless SRAs were bound to replicate like maggots upon the cadaver of its host, we might well miss them even if they had landed safely on the Earth. After all, we sometimes fail to find airplanes and ships in distress even when we are looking desperately for them. A machine that made only a few copies of itself might be unobtrusive, or it might have come a few million years ago, or it might not have arrived yet. And even if it has arrived, as Ronald Bracewell suggested many years ago, it may be waiting for the right moment to make its presence known.[36] The ability to get a machine to every planetary system eventually is different from the ability to have a machine on every planetary system at any one time.

Besides, if the SRAs were to reproduce without restraint, they would become a rot on the galaxy eventually, and then we could not miss them. But it is not clear why a civilization should wish to create such a nuisance. It is even less clear why a civilization should feel *compelled* to plague other worlds so. And once again, an impossibility proof is not worth much when its conclusions are not inevitable, and when its crucial assumptions are highly questionable, perhaps even far-fetched.

Nevertheless, the examination of the future conjured up from von Neumann's idea has led to a clearer understanding of life and exploration. And the future painted by Sagan may still guide our way into the cosmos. Not unlike science fiction, these two scientific scenarios force us to contemplate possibilities and teach us how to rehearse in the imagination steps that our descendants might need to take some day.

Fermi's Question and the Exploration of the Galaxy by Living Beings

As we discussed in Chapter 7, fusion rockets, ramjets driven by nuclear catalytic engines, and even superluminal starships are consistent with current physical theory. Surely, then, an advanced civilization could attain a velocity one-hundredth that of light, and thus it would take it about ten million years to arrive at the furthest confines of the galaxy. Therefore, if technological civilizations are as prevalent as the proponents of SETI would have it, some of them would be billions of years old now, and so ETIs should have been here already.

There is no question that interstellar travel at that velocity would pose a variety of social difficulties for us. Chief among them is that it would take 400 years to arrive at the nearest star, which means that the trip must be completed by the descendants of the original astronauts, but we are not sure that we can entrust the success of the exploration to people that would not have been born when the decision to explore the galaxy is made.

Nevertheless, we can cook up several scenarios in which the social obstacles are overcome as a species begins to migrate to the stars: an authoritarian regime forces the issue, or there is forewarning of a cosmic catastrophe, or the migration is simply a natural consequence of a long and massive colonization of the species' own planetary system. In such a case, Fermi's impossibility proof seems plausible again.

It does no good to reply with reasons why a civilization may not choose to explore the galaxy. Given that we are average, according to the Principle of Mediocrity, and given that we are an expansionist species, we should expect some of those many alien civilizations to be expansionist as well. But where are they?

We could, however, find excuses for why we have not *detected* an alien presence in our solar system. It would be enormously difficult to spot even a large starship that came within a few astronomical units from Earth, a distance that may be quite suitable for an alien species to conduct a survey of our solar system. At that distance it is not easy to detect asteroids smaller than a kilometer across, even when we are searching for them. Also, a starship may come in no closer than Saturn but send much smaller probes into orbit around the other planets. Their advanced stealth systems may be beyond our technological ability to detect. Or the ship may have been here already and gone home (or gone silent). The excuses may be limited only by our imagination. But is it plausible to ascribe such motives to alien species?

The Wisdom of Contact

To make matters worse for SETI, the Principle of Mediocrity should lead us to question the wisdom of trying to communicate. It is clear that in our expansion in our own planet we have done our best to eliminate all significant competition from other species. The last thing we wish to do is advertise our presence to more advanced species who may then wish to occupy our niche, and in the process may need to get rid of the local pests, or at least bring them under control. In some circles there is the feeling that advanced creatures must somehow be wise and benevolent, although under the guidance of the Principle of Mediocrity it would be difficult to see why. In the first place we have a history of ruthlessness toward species that become obstacles to our aims; we have been ruthless even to other human cultures. Consider, for example, as Ron Bracewell has pointed out, what the response of suburbanites might be if raccoons became much smarter. They would be such pests that suburbanites would go to great lengths to wipe them out.[37] And few of us would lose much sleep over that. Indeed, when we try to poison cockroaches and rats, or hunt the coyotes that prey on our sheep, the issue of benevolence or malevolence seldom comes up.

In light of these considerations, some suggest that we should lay low until we are in a better position to do battle, if need be. There are others who argue that the issue is moot since we have been radiating into space our radio and television signals for a long time. That may be so, but those signals would be very weak and well scrambled by the time they leave the solar system, and in view of the difficulties we have recognized in trying to look for alien transmissions, and the low powers of most of our own transmissions, it is not unreasonable to suppose that detecting life on Earth from, say, one hundred light years away would involve a rather substantial amount of luck. Of course, some transmissions, radar beams for example, are very powerful. And in any event, our great activity in the radio range alone might indicate to another civilization that a relatively advanced technology exists here. Frank Drake and Carl Sagan sent a message in 1974 using the Arecibo radio telescope. It provided information about humankind in binary code. Others have sent radio messages as well.

The Principle of Mediocrity

Sagan proposed the Principle of Mediocrity to bolster the justification of SETI's program, which is to listen to the universe with radio telescopes in the hope that an artificial combination of pulses may be identified, to decipher such a signal, and perhaps to respond, thereby initiating the most extraordinary communication in the history of the human species. Surely, if we are average, there may well be a whole club of civilizations out there, and with just a little effort we might be able to join them. But then SETI's opponents also assume the principle to bolster their impossibility arguments à la Fermi.

The Principle of Mediocrity is prompted by the notion that our belief that the Earth was the center of the universe, and that we were thus the pinnacle of creation, sprung from some primitive anthropocentric view of the world later reinforced by religion. Remove the notion of humans at the center of things, and it becomes imperative to face up to our average nature. Nevertheless, our ancestors did have good reasons for thinking that the Earth was the center of the universe. It took a lot of ingenuity, genius, to overcome devastating objections to the idea of the motion of the Earth (see the treatment of the Tower Argument in Chapter 3). Nor did they think that the Earth was at the center of the universe because it was special in any commendable way. On the contrary, the heavens were eternal, and unchanging, our example of perfection. Change and corruption could take place only in the lowly Earth. Copernicus himself resurrected the Pythagorean claim that the Sun should be at the center of the universe since it was obviously so much nobler a body than the Earth. Thus, Sagan's account does not live up to historical examination. It does not follow from the Copernican Revolution that we are average.

Moreover, although we have now reason to believe that the Sun is an average star and that the Earth is not the center of the universe, we cannot say that we have similar reasons about our own standing in the realm of life. In the one relevant aspect—intelligence—we are clearly not average in the domain that we have been able to observe.

SETI should therefore be discussed without the burden of the Principle of Mediocrity, for or against. No arguments should assume, for example, that, if there are any ETIs, they should be so strikingly similar to us that we can make reliable, quasi-probabilistic guesses about them based on intuitions about ourselves. Future arguments must include a wide range of considerations from biology and space science, as we will see later.

SETI enthusiasts are wont to make rosy estimates about the probability of life. In one key factor, the number of planets in the galaxy, the finding of thousands of exo-planets smiles favorably upon their enthusiasm.

Some proponents of SETI claim to use "subjective probability" in arriving at those estimates. According to T. Fine, the subjective interpretation of probability "maintains that probability statements are derived through a largely unassisted process of introspection and are then applied to the selection of optimal decisions or acts."[38] Furthermore, this subjective view "encourages the holder to fully use his informal judgment, beliefs, experience in arriving at probability estimates."[39] Although personal, such estimates are presumably not arbitrary because "there are reasonable axioms of internal consistency between assessments and constraints that force the user to learn from experience in a reasonably explicit way."[40]

This view of probability, together with the Principle of Mediocrity, has indeed encouraged some SETI enthusiasts to make highly optimistic pronouncements about the likelihood of planets with life, intelligence, and technological civilizations, based on the fact that the Earth has all three. But can these scientists justify what amounts to giving a statistical distribution from only one case?

The intuition behind subjective probability is that a scientist who has already learned from experience, and who is in a situation to which his expertise is relevant, may come up with reliable hunches as to what is the right action to take. Indeed, we may measure such probability by determining how much he is willing to bet on a course of action over its alternatives. But this notion of probability does not apply in the case of SETI. On this subject we have learned nothing from experience because we have had no experience to learn from; nor can we use our expertise about the Earth because our theories are not yet developed enough to make decent guesses about how representative the Earth is. In the years ahead we might, if we continue to increase the sophistication of telescopes in orbit. But right now, estimates about how many planets have life and so on are arbitrary. And the fact that future exploration may force us to "learn from experience" does not make them any less arbitrary now. Using our "informal judgment" and "belief" when our ignorance is so vast does not seem any more encouraging.

A related misuse of probability comes in the practice of splitting the difference. The optimist will use his subjective probability to estimate that in every mature planetary system there will be at least one planet with life (the probability of life is one); the pessimist will say that the probability is zero because life could have arisen only on Earth. And then there are those congenial types who declare that the truth must fall somewhere in between, and so decide that a probability of one-half (or one-fourth or one-sixteenth) is a "conservative" or "reasonable" estimate.

Imagine, however, that I am given a photograph of a building that could be either Fort Knox or an empty warehouse, and that I am asked to estimate how much wealth that building contains. Suppose that I know that there are $200 billion in gold in Fort Knox. And now, since I have no idea which building it is, I split the difference and estimate that there are $100 billion in it. Whichever building it turns out to be, my estimate will be off by $100 billion, not a small mistake. In the case of ETI, our estimates of probability should be based on our knowledge of the universe, not on reaching a compromise between uneducated guesses. As space science advances, we will have more insightful things to say about the chances for extraterrestrial life. For ETI we will have to take a few additional steps.

Extraterrestrial and Human Science

Travel teaches us not only about other places and people but also about ourselves.[41] Likewise trying to understand what other intelligent life might be like teaches us about our own intelligence. And trying to understand how alien intelligence may view nature teaches us about what our own views of nature amount to. In SETI we find almost bare many common assumptions about the origin, development, and nature of science. Thus, from an analysis of SETI we may be able to draw some interesting philosophical lessons.[42]

In this section I will be concerned mainly with three notions that are frequently advanced by SETI proponents. The first notion is that once life appears on a planet, intelligent life is also very likely. The second is that once intelligence appears on a planet, science itself is likely. The third is that all scientific civilizations have something in common (i.e., an overlap in their scientific views of the world) and thus the basis for the beginning of communication between them.[43] The first two notions are advanced to support the contention that there is probably someone to look for. The third gives us hope that contact, if we make any, will be productive.

In spite of their initial plausibility, I will argue that these notions are plagued with less than obvious assumptions at many levels, and that they lead to a questionable account of our views of nature.

SETI proponents believe that life can begin elsewhere, that once it begins it is likely to become more complex, and that complexity produces intelligence. Presumably, as time goes on, intelligence will improve its attempts to understand what the world is like—thus begins the almost inevitable road to a technological civilization. Whether life can begin elsewhere is a matter of ·great controversy, as we have seen. But I will grant for the sake of argument that it could. In the same spirit I will grant that, at least for some time, the complexity of life may increase; and I will also grant that intelligence is the result of certain complex biological organizations. But granting all these crucial assumptions of SETI is not the same as granting that alien technological civilizations are very likely.

Let me take stock of what I have granted. After some primitive form of life appears on a planet, it will not remain uniform for long. Small variations in the environment and other factors will bring about diversity. Of course, diversity is not the same as complexity, but it gets us on the road to it, for diversity means that there will be different kinds of biological structures and different ways of interacting with the environment. And the possibility then arises that eventually some of these structures, say A and B, will combine, thus improving the chances that a new structure C will arise to coordinate their work. And now A, B, and C together will form a new whole that is more complex than either A or B as separate individuals. During the first couple of billion years of life on Earth, prokaryote cells were the most prevalent, perhaps the only, form of life. These cells in which the chromosomes are not protected inside a membrane (the nucleus) eventually led to cells with nuclei (eukaryotes), which are more complex. According to Lynn Margulis, this important step came about by the symbiosis of different kinds of prokaryote cells.[44] In any event, once cells with nuclei appeared it was possible to form organisms that combine many of these cells, sometimes billions of them. These organisms are very complex wholes of eukaryote cells that perform many different but coordinated functions. Although after billions of years the increase in the complexity of life can be considerable, complexity is not always bound to increase with time. Changes in the environment of a planet, some of them caused by life itself, may make it very difficult for all but simple organisms to survive on that planet.

Let me concentrate now on a particularly interesting kind of complexity. Eventually some Earth animals developed intricate patterns of muscles and bones so they could move about, external senses to give them information about the world, and internal senses to monitor a variety of organs. It does not take much to see the advantage of coordinating these functions. A successful predator not only sees the prey but also can move to catch it. On Earth, a popular answer to this problem of coordination is the central nervous system. And this is an interesting answer because it is in connection with a highly complex central nervous system that intelligence becomes conspicuous.

A highly complex central nervous system is not limited to just one way of handling the information that it receives from the world: it can rout and combine information in a variety of ways; it can compare sense modalities; it can store information and consider alternative actions; that is, it can make use of memory and imagination.

Visual perception is a good illustration of this kind of complexity. At a very primitive level we may suppose that the detection of light is enough for a certain organism in order to move toward or avoid the light. The next step comes when the organism gains an advantage by being able to discriminate visually between objects, which may be achieved by making internal representations of those objects. These representations grow in sophistication, and the corresponding nervous structures in complexity, when the "input" from the eyes is coordinated with that from other senses. For example, when we are looking at a painting of a group of people, our eyes are not stationary. First, the eye muscles make the eyes scan continuously. Second, our heads move sideways as well as up and down. Our whole bodies may also move, carrying our heads, and thus our eyes along. But the images of those people remain stationary because the brain takes into account the automatic movements of the eyes, as well as our body position, as it receives information from the inner ear, which keeps track of the inclination of the body with respect to the Earth's gravitational lines of force, and from hundreds of skeletal muscles.

Visual perception is also easily affected by the other senses. As we walk down a dark street at night, we may perceive what appear to be some bundles a few steps ahead. But one of those bundles suddenly becomes a sharp image when we hear the distinct growling of a guard dog. Perception also takes into account memory and imagination. An artist well trained in the history of art may see many more details in the painting and many more relationships between different elements of the painting than most of us can, just as a well-trained naturalist can detect a rare bird in a bush where most of us can see only foliage.

The more complex the central nervous system, the more complex the relationship between the organism and the environment, for the organism gains more degrees of freedom. Thus, intelligence arises out of perception and other biological structures as the complexity of those structures increases. This account agrees with Piaget's description of intelligence as an instrument of adaptation not necessarily tied to the immediate and momentary demands of the environment (human beings, for example, can figure out solutions to problems that will confront them far away and years hence).

Let me sketch now the main hurdles that life has to overcome on its way to becoming an advanced technological civilization. To say that intelligence is adaptive is to say that a highly complex central nervous system (or its equivalent) is adaptive. But then intelligence is adaptive only for certain kinds of organisms and not for others. It would be adaptive for primates, for example, but not for cockroaches. Let me illustrate the point by means of an analogy. It is well known that the opposable thumb is a highly adaptive feature of human beings. But it would not be so for horses. And it does not even make any sense to ask whether it would be for cockroaches, since roaches do not have the kinds of physical structures to which opposable thumbs can be attached.

We might think that roaches would be better off if they were smarter. But to put the point properly we have to consider whether roaches would be better off with more complex brains. And now we may begin to see the difficulty: there is a price to pay all along the way to intelligence. The price is that a complex brain demands a high metabolism. In a minor way the same point may be made about sight, which also seems to be quite an advantage. Imagine that a population of small mammals has come to live in dark caves. The brain structures of sight use a lot of energy, and so these mammals have to spend much time and work getting that energy. Since sight is of marginal advantage in the dark caves, the mammals that preserve sight are not as competitive as others that use only a fraction of the energy to enjoy improved hearing, touch, and smell. It would be nice to have sight, but a mammal of that size can't afford the price to keep it. And for a population of sightless mammals, it would make no sense to develop it.

Likewise, an increase in the complexity of the brain requires that the organisms of the species in question gain some advantages that compensate for the price in metabolism that they have to pay. In the case of many species on Earth, including ours, those advantages have been there. But we should not expect that they would be there on any other planet where life may evolve.

Consider our kind of intelligence: mammalian intelligence. If the dinosaurs had not become extinct, mammals would have remained small vermin. Large mammals could not evolve because an increase in size would make it easier for dinosaurs to prey on them. But the price that mammals would have to pay for a bigger and more complex brain would probably be a bigger body. Dinosaurs, thus, precluded the evolution of high intelligence in mammals. Today we ourselves are a cap on the evolution of high intelligence by others. Suppose once again that raccoons become increasingly intelligent. As it was pointed out earlier, they would become such pests that humans would probably hunt them to extinction. Our very way of life tends to wipe out animals that enter into close competition with us.

Let us imagine a planet very similar to our own. Let us suppose that in that planet also the conquest of the land by fish would have provided the necessary opportunities for an increase in the complexity of the brain. But let us also suppose that on that planet insects had already appeared on the land and were even more successful than on Earth. Because of their physical constitution, insects are not likely to grow large enough to develop the sort of large brain associated with intelligence. But insects have many adaptive features that serve them quite well. Thus, they can be successful without being smart. In that planet they rule the land: any fish that crawls out of the water will be eaten by insects, and if perchance eggs from that species are not only laid but hatched, the young fishes will be devoured. Intelligence as we know it is not likely to arise. The smartest being on that planet would be some kind of octopus (it does no good to point to whales and dolphins—those are mammals and would have never evolved if vertebrates had not developed on land to begin with).

On other planets the cap may come from many different kinds of beings, even if their own intelligence is rather modest by our standards. All it takes is that in some other respects they can adapt *first* to the land, or whatever key environment we consider. But what enters into that timing? Most often just accidents of natural history. For example, it is possible that the disappearance of the dinosaurs may be traced in large measure to the collision of a gigantic asteroid with the Earth. But there is no guarantee, let alone a law of nature, that accidents of natural history are going to favor the development of high intelligence.

Let me imagine, nonetheless, that on some planets central nervous systems as complex as ours, or even more complex, do evolve. Will technological civilization then come about? Not automatically. It has to be the right kind of intelligence: technological intelligence. The evolution of *human* intelligence is tied to the use of tools for hunting and many other purposes. But the evolution of this mode of interaction with the world makes sense only if you have the right kind of body. Dolphins, for example, which are creatures with complex brains and perhaps high intelligence (even if not in our class), have no hands, to say nothing of opposable thumbs. There is a clear sense in which we express our intelligence by having the appropriate bodily interaction with the environment. A technological intelligence would not be adaptive unless the right kind of body developed along with it. Spears may have been a sensible option for our ancestors, but harpoons would not have been so sensible an option for the ancestors of dolphins.

Nevertheless, let me suppose that *technological intelligence* does arise and takes over a planet. *Technological civilization* still does not follow automatically. A technological civilization is in part the result of complex social processes; thus, the required type of intelligence must be not only technological but also social. Nevertheless, even if we have the evolution of this kind of intelligence, a *highly advanced* technological civilization may not arise. One reason is that high technology may well require the development of science. On our own planet a turning point came when the new science, culminating in Newton, was able to bring together astronomy and physics. Yet in a planet very similar to ours but perennially covered by clouds (or in a solar system traveling through a dust cloud) a comparable development of astronomy would be most unlikely.

Imagine, though, that we have a favorable physical environment where intelligent beings (both socially and technologically) can receive the inspiration and rewards that would take them on scientific paths blessed with the right kinds of intellectual breaks. We still cannot expect an advanced technological civilization. For having the right physical environment is not enough. Social factors may still prevent the development of science as we know it (let alone a more advanced science). It is plausible to suppose that the progress of science requires that ideas may be criticized and that alternative conceptions of the world be developed and defended even when the majority disagrees with them. But in a species biologically inclined to a degree of social cohesion greater than ours, the criticism of the metaphysics of the society (e.g., of their account of the origin and nature of the world) may be seen, or felt, as a threat to the cohesion of the society and put down at once. It seems that in our world, science barely made it; on that other planet, science would have no chance.

I do not wish to argue that a technological civilization could not arise on a different planet. My intent is merely to point out that the process is by no means automatic, that it requires many good breaks from natural history.[45] A critic may argue that natural selection could have gotten around most if not all of the obstacles I have mentioned. All it takes is a bit of imagination, and we know how imaginative natural selection can be. For example, one of the reasons why advanced technology seems to need a social milieu to exist is that no one human being can fully develop a theory as comprehensive as, say, Newtonian mechanics (it took centuries), to say nothing of all the other branches of physics, chemistry, and so on. But within one school of thought it is difficult enough to come up with a few good ideas. To be able to see their flaws, possible means of improvement, or their connections to other areas of science often requires that we look at those ideas from many different points of view.[46] One human being could not do all this. Science and advanced technology require a division of labor.

Imagine, however, a planet on which a single organism—not a single species, a single organism—comes to dominate even more than human beings do on Earth. This would be a strange organism that covers the environment like a comforter and grows larger by creating more branches of itself until it has finally covered much of the planet (if the food supply decreases, this intelligent organism will either "farm" differently or drop off a few branches). Instead of a central brain, this organism has something that rather resembles a network of ganglia (large, complex ganglia to be sure). Although the action of the ganglia tends to be coordinated, in a network that large there must also exist a fair degree of decentralization. In that case, ideas may be brought up by one particular ganglion and criticized by other ganglia, and so on. The concept of self of this organism may be quite different from ours, but the point is that in a single organism we may find the equivalent of a whole species. So this organism could develop an advanced technology, even though, strictly speaking, it is not really social.

It is clear, then, that if certain avenues of development are closed to life, natural selection may find others. But the price of alternative natural histories would be alternative forms of intelligence and eventually alternative ways of formulating views of the world. The reason is that the brains (or their equivalent) that would result from such radically different natural histories would arise from entirely different biological structures, and thus, in coordinating these structures, the developing brain would face different evolutionary problems and would have different solutions and opportunities at hand. In a hospital's neurological ward, we find people whose brain structures have been altered and who thus have peculiar ways of perceiving and conceiving of the world. Of course, their modes of thought are maladaptive, just as skeletal structures that deviate from the norm may be maladaptive. But for different creatures, different brain structures and their corresponding modes of thought may be as adaptive as their different skeletal structures are. The consequence of this point is that the science of a species, or kind of organism, may be relative to its natural and social history. Thus, species with very different natural histories may have little overlap in their scientific views of the world. If this is so, there would be much less in common to serve as the basis for interstellar communication with other technological civilizations than the proponents of SETI make it out to be.

Defenders of the SETI program often assume that advanced sciences and technologies must exhibit a high degree of convergence. The grounds for this assumption are presumably that, as science grows in scope, the brains that produce that science must reckon with all-pervasive features of the universe. Just as dolphins and fishes have very different evolutionary histories but similar shapes because they both live in water, so sciences that deal successfully with the basic forces of the universe must come to similar views. Nature presumably already offers many cases of convergence: placental and marsupial wolves, and camera eyes in squids and mammals, to mention only two of the most striking.[47] Furthermore, when it comes to communication with advanced technological civilizations, we are talking about species that at a bare minimum have invented means of electromagnetic transmission and may also have embarked on a program of space exploration. Their views may be superior to ours (having been around longer), but, surely, they must overlap with ours to some extent, for at least to some extent they and we are successfully applying the laws of electromagnetism.

Nevertheless, the matter is not this straightforward. We must realize that even all-pervasive features of the universe would be interpreted differently by different scientific intelligences. As we have seen, a highly complex brain can deal with the environment in a very flexible and indirect manner. Moreover, it is not one brain but an ensemble of brains in very complex social relations that deal with the universe through science and technology. Whereas in the case of the ocean we had direct pressure (selection) on aquatic animals, in the case of the deep forces of nature we have many different ways of handling the pressure (indeed, a double tier of evolutionary slack). Even in the case of the ocean, animals with very different evolutionary histories have different shapes, as we can tell just by looking at crabs and salmon (fishes and dolphins are much more closely related). The very same "feature" of an environment impinges very differently on different organisms. A hot spring may kill some fish while making bacteria thrive. It is a mistake, therefore, to describe the situation as if different brains were dealing with the same problems. We rather have different brains dealing with different problems. Indeed, those different brains will have (1) different starting points for inspiration, (2) different motivations, and (3) different social means of dealing with conceptual matters.

As for the overlap in electromagnetic theory, we should guard against confusing an overlap in *performance* with an overlap in *content*. For in a limited domain two radically different views may allow us to do pretty much the same. As a guide to navigation, the astronomy of the ancients was not surpassed by the astronomy of Copernicus and Newton until long after Newton's death; and it remained competitive until the advent of recent technology. But according to the ancient view, the immobile Earth sat at the center of the universe while the stars were fixed on a gigantic sphere that rotated around the Earth. By keeping the stars in that sphere, it was possible to calculate very precisely their position in the sky at any time of the year. And by reference to that position a sailor or an explorer could chart his course. In many respects it is still easier to apply the ancient view. In any event, to some extent the ancient and the modern views give us very similar practical guidance; they allow us, in a limited context, similar performances. But the views are not only different, they contradict each other: one forbids the motion of the Earth around the Sun; the other requires it. If perchance we receive electromagnetic transmissions from another species, we should not conclude that those beings must have the equivalent of Maxwell's laws of electromagnetism. We may need Maxwell's laws in order to describe, to ourselves, what those beings do. But their actual "laws," if they even think in such terms, may not be any more equivalent to Maxwell's than the Greeks' lack of motion of the Earth is equivalent to Copernicus's motion of the Earth around the Sun.

We see then that there is no inevitable, nor highly probable, connection between the appearance of life and that of intelligence; nor between the appearance of intelligence and that of an advanced technological civilization.

None of the preceding rules out the possibility that extraterrestrial science exists. My aim has been to investigate the assumptions behind the optimism prevalent in SETI in the context of natural and social history. That investigation allows us to determine some of the conditions that make *human* science possible.

Ethical Obligations to Intelligent Alien Life

It seems intuitive that in the case of intelligent alien life we are even more likely to have ethical obligations, well beyond the instrumental case made for alien bacteria in Chapter 6. Obviously, the biological significance of highly complex beings will be greater still, although the key issue here is the fact that they are intelligent, presuming that we can make that determination. Still, it seems, in any event, that biological considerations about the nature of intelligence and morality will prove of great importance. Charles Darwin explained how the foundation of morality "lies in the social instincts, including under this title the family ties."[48] Developing this and other evolutionary insights, Patricia Churchland[49] and several other authors have argued that the morality of species is indeed influenced by their biology. Such a point of view may perhaps present a difficulty for the intuition that intelligent life will demand a greater ethical obligation, for intelligent aliens would most likely be the result of a different basic biology: do they have DNA, for example? And if they do and form cells, do they have the equivalent of prokaryotes and eukaryotes? And even if they did, it would be astonishing if their evolutionary history were anything like ours, for they are unlikely to have gone through the same sequence of environments that we have. Thus, their equivalents to nervous systems would be extremely unlikely to resemble ours. As Michael Braide points out, we would not have a morality in common with them.[50] Some may go further than Bradie and argue that, even if the intelligent aliens were moral agents in some way, we might not even be able to recognize them as such.

This is not the place to discuss whether the biological approach to morality is correct. The reader may consult my work (1998) for arguments to that effect, which show, for example, that objections based on the "naturalistic fallacy" are actually instances of very poor reasoning.[51]

Moving along a more optimistic avenue of thought, Peter Singer (1981) argues that social animals restrain their behavior to one another and do things for each other.[52] Presuming that many of the intelligent aliens we might meet are social animals, this characteristic would give us something important in common. Nevertheless, even if we grant Singer's point, we must realize that evolution produces very different ways of being social: take humans and bees as a clear example. Even closely related species, such as chimpanzees and bonobos, are social in different ways because of slight anatomical differences, as Dale Peterson's examples make clear.[53] Extraordinary biological differences could be expected to result in extraordinary differences in being social, and, presumably in being moral.

Julia Sandra Bernal explains how bipedalism, by freeing human hands, led to greater development of tools and access to many sorts of foods, while creating incentives for much higher cooperation and social expectations within human groups.[54] As human intelligence increased, so did the size of the newborn heads, and thus the need for human babies to be born prematurely, compared to other species. This in turn led to many social mechanisms to support the new mothers and their infant children, including, Bernal argues, a tendency toward (serial) monogamy and strong emotional ties on the part of the fathers. She then argues that these biological factors played a great part in the development of morality and law among hunter-gatherers and the much larger human societies that appeared later. The chances of an alien intelligent species to evolve in a similar way to us are simply extraordinarily small, and thus they would be unlikely to have a similar morality (or a similar range of moralities) to ours. This would be in addition to having different cognitive evolutions, which would probably make communication very difficult as well.

According to E. O. Wilson, moral intuition seems to be the final moral arbiter.[55] What exactly would prevent the intelligent aliens from having similar moral intuitions to ours? The answer is implied in the previous remarks: moral intuitions are based on moral emotions that result from the operation of certain brain regions. That operation, however, is the result of a long evolution (as are those brain regions), and, therefore, *radically different intelligent species are likely to have very different moralities*. To see that such differences are not limited to mere rules of conduct but apply to ultimate moral principles, let us make use of John Rawls's Veil of Ignorance.[56]

Rational agents, Rawls tells us, choose consistently with their best interests. Looking through the Veil of Ignorance (unaware of his or her social position and personal characteristics), a rational agent would choose that the society should allow him or her maximum freedom consistent with equal freedom for others. Such rational agents, Rawls believes, would also reject the Principle of Utility, since this principle requires the sacrifice of individual happiness in order to achieve the greatest balance of happiness over unhappiness. If I, or anyone else, were to agree to the Utilitarian arrangement, I may be acting against my own interests (and thus being irrational) because the unfair sacrifice may fall upon my shoulders. As Rawls points out, this cuts too much against the grain of human nature.

If Rawls's famous account is correct, then radical differences in biology may alter the results of deliberations under the Veil of Ignorance. Imagine that rational ants exist in some faraway planet. An integral part of being an ant is being dominated by a group mentality. An ant's equivalent of Rawls's *Theory of Justice* may find abhorrent Rawls's First Principle of Justice (of equal freedom), while looking favorably upon the Principle of Utility. And, of course, the biology of many intelligent aliens is likely to be even more radically different from ours than that of rational ants.

In a clear sense, however, the problem seems to be that the moral values of a species are relative to its biology. But this is a problem with a solution, according to Singer, as long as we treat the relative values of different intelligent species with impartiality. Were we to encounter intelligent aliens, we would not wish to have them impose on us their notions of proper behavior regardless of our interests. This obvious realization should guide us in the approach we should take toward them, as we will see again in Chapter 10. Where the only relevant difference between our wishes and theirs is that they are ours, we would generally not be in a position to give them reasons why they ought to behave as we want them to. Intelligent beings should presumably be able to come to the same realization. To insist on ignoring impartiality would be done in the knowledge that they have no rational claim upon the behavior of other intelligent species with which they are interacting. As we will see in Chapter 10, "objective" values, then, need not be necessary for an ethics that expands to many societies, let alone to alien intelligent species. This view will be reinforced there by the views of J. L. Mackie.

We do have an element of universality in Singer's account based on relative values, namely, once again, the realization that "one's own interests are one among many sets of interests, no more important than the similar interests of others" (p. 106). Singer concludes that "Wherever there are rational, social beings, whether on earth on in some remote galaxy, we could expect their standards of conduct to tend toward impartiality, as ours have" (p. 106).

In addition to the instrumental case for ethical obligations to preserve alien life, if that life happens to be intelligent, we would have to consider what sort of behavior we should exhibit toward those aliens, just as we would need to worry about how they might treat us. It is extremely likely that finding ways of getting along will be crucial for all the species involved. Singer gives us the key to what we owe them: to treat them according to his Principle of Impartiality. This is what Reason would demand from us, as it would demand from them.

Apart from the likely difficulties in communication, lacking common-ality of interests may obstruct attempts to behave ethically toward each other. Nevertheless, a reason for optimism is that in new circumstances complex intelligent beings are capable of developing new interests. Of course, we have no guarantee that Singer's theory will be born out in every case: we may be unlucky enough to encounter the Nazi equivalent of a species. Still, Singer's view on impartiality gives us some hope in facing a future that may await us in the stars.

Can SETI Be Justified?

Is SETI a waste of time and money? I do not think so. SETI does at least two valuable things. First, it provides an extraordinary opportunity for a shortcut in our search for life in the universe. For obviously, if we detect intelligent civilizations, we will have settled the issue of the possibility of extraterrestrial life, which otherwise may take hundreds, perhaps thousands, of years to re-solve, if it can be resolved at all by space travel.

SETI's chances of success may be slim, but if we do succeed, the results would be of the greatest significance. SETI is like a lottery ticket: as long as the investment is small, we have little to lose and much to gain.

Second, SETI provides special motivation and in some cases inspiration for many researchers who work in areas related, however indirectly, to the issue of the origin and the evolution of life. Indeed, to be fair to the SETI enthusiasts, much of their work has concentrated on improving our knowl-edge of several of the links in the chain between the origin of the galaxy and the origin of life.

Whether SETI succeeds or not, however, I suspect that its main possible contribution lies elsewhere. Just as exobiology can provide a very useful con-text in which to ask questions about the origin and evolution of life, SETI may become a useful framework to examine the nature of our intelligence and our technological civilization.

Perhaps a few readers expected a discussion of the evidence of ETIs pro-vided by the many sightings of unidentified flying objects (UFOs) and their prodigious feats. Over several decades I have noticed that such sightings are eventually explained away as weather balloons, or something else of terres-trial character, only to be replaced by a new batch eventually destined to the same fate. Ninety-nine percent of the cases that were brought to my attention, it seemed. It is difficult not to become jaded. The most recent episodes received great press about lights in the sky displaying physics that was just plain impossible for human creations. Well, as I mentioned earlier, I have seen Jupiter, shining through low distant fog in the early evening, move at "extraordinary speeds" and "turn around" in extremely sharp angles. Even "more impossible physics" than those recent "spaceships" showed. And it is also something that happens to points of light in a dark background.

In the case of the most recent "UFOs," an analysis of the videos shows, according to Mick West, that the reported "glowing aura" is just an artifact of thermal cameras; the "impossible accelerations" were caused by sudden movements of the camera; and the video of the "impossible rotations" shows that when the "UFO" rotated, other patches of light in the scene also rotated, which can best be explained by a rotation of the camera. The rest of the videos were, well, similarly embarrassing.[57]

Concerning the issue of evidence about past visits by aliens, the reader may do well to consult David Lamb's excellent book on this aspect of SETI.[58]

Notes

1. Carl Sagan (ed.), *Communication with Extraterrestrial Intelligence* (MIT Press, 1973).
2. For a description of Cyclops, see Bernard Oliver's description in Sagan, *Communication with Extraterrestrial Intelligence*, pp. 279–301. The report on the project was published by NASA: CR 11445.
3. An earlier version of this section was published as Ch. 15 of my *Evolution and the Naked Truth* (Ashgate, 1998).
4. Carl Sagan, *Pale Blue Dot* (Random House, 1994), pp. 39, 372.
5. Authors more contemporary than Fermi have further developed the argument criticized here. See, for example, Frank J. Tipler, "Extraterrestrial Beings Do Not Exist," *Physics Today* 34 (April 1981): 9–38. Sagan twice blocked the publication of this paper. In turn, Senator Proxmire used it to block Sagan's ambitious SETI project: Cyclops. For recent commentary (and the reference to Fermi), see Paul Davies, *Are We Alone? Philosophical Implications of the Discovery of Extraterrestrial Life* (Basic Books, 1995), based on his series of lectures at the University of Milan in 1993.
6. John von Neumann, *Theory of Self-Reproducing Automata*, ed. A.W. Burks (University of Illinois Press, 1966).
7. *Advanced Automation for Space Missions* (NASA, 1982).
8. The exceptions would present a variety of inconveniences: Europa is covered by a layer of ice and presumably an ocean of water. Io has a very unstable surface. And so on.
9. As noted in the previous chapter, the classic critique is Hubert Dreyfus, *What Computers Can't Do* (Harper & Row, 1972), revised and updated in 1979.

10. Connectionist approaches, which Dreyfus does not analyze, and which present a significant alternative to von Neumann's view, are better able to handle context. See an optimistic treatment such as Paul M. Churchland's, *A Neurocomputational Perspective: The Nature of Mind and the Structure of Science* (MIT Press, 1989). I find Churchland's view very plausible, and this paper supports it by undermining the view of mind (and body) put forward by von Neumann.

11. An equivalent situation occurs when living things move into new environments to which their adaptation may leave much to be desired. That would not be the case for the situation of the SRAs.

12. For a very interesting illustration see Gary Marcus, *The Birth of the Mind: How a Tiny Number of Genes Creates the Complexities of Human Thought* (Basic Books, 2004).

13. Gunther S. Stent, "Strength and Weakness of the Genetic Approach to the Development of the Nervous System," *Annual Review of Neuroscience* 4 (1981): 163–194, p. 187.

14. Stent, "Strength and Weakness of the Genetic Approach," p. 188.

15. Stent, "Strength and Weakness of the Genetic Approach," p. 188.

16. The Protein Data Bank model data for Figure 8.2 is derived from J. S. Kavanaugh, P. H. Rogers, and A. Arnone, "T-to-T High Quaternary Transitions in Human Hemoglobin: desArg141alpha Deoxy Low Salt." http://dx.doi.org/10.2210/pdb1 y0d/pdb

17. Simon LeVay, *The Sexual Brain* (MIT Press, 1993), p. 89.

18. LeVay, *The Sexual Brain*, p. 92.

19. LeVay, *The Sexual Brain*, p. 93.

20. Robert A. Freitas Jr., "A Self-Reproducing Interstellar Probe," *Journal of the British Interplanetary Society* 33 (1980): 251–264.

21. Alan Bond (ed.), *Project Daedalus: The Final Report on the BIS Starship Study* (British Interplanetary Society, 1978). Daedalus would have been 190 meters long and would have weighed over 50,000 metric tons.

22. We have already found hundreds of extraterrestrial gas giants, some even larger than Jupiter. In some such planets, we could presumably expect moons even larger than Mercury, similar to Jupiter's.

23. See, for example, the references earlier to Dreyfus and Churchland.

24. This and many other illustrations are explained by Churchland.

25. This is the interpretation found in Thomas S. Kuhn, *The Structure of Scientific Revolutions* (Chicago University Press, 1970).

26. Antonio Damasio, *Descartes Error* (Putnam, 1994).

27. Gonzalo Munévar, "Human and Extraterrestrial Science," *Evolution and the Naked Truth*, Ch. 2.

28. Quoted in many websites, including www.quotationspage.com/quote/255.html

29. Richard S. Smalley, "Of Chemistry, Love and Nanobots," *Scientific American* 285, no. 3 (September 2001): 76–77.

30. George M. Whitesides, "The Once and Future Nanomachine," *Scientific American* 285, no. 3 (September 2001): 78–83.

31. Steven Ashley, "Nanobot Construction Crews," *Scientific American* 285, no. 3 (September 2001): 84–85. In the same issue Drexler had an opportunity to address the content of the objections, but instead simply suggested that his critics were unqualified and that successful research in the field favored his idea. Ashley's article suggests otherwise. Eric Drexler's proposals can be found in his *Engines of Creation: The Coming Era of Nanotechnology* (Anchor Books, 1986) and *Nanosystems: Molecular Machinery, Manufacturing and Computation* (John Wiley & Sons, 1992).

32. Under very special laboratory conditions, DNA can be used to make nano-machines, but nothing of the sort relevant to our discussion.

33. None of the foregoing shows, incidentally, that artificial life is impossible. It would be *life*, even if made from scratch in a laboratory or a factory, as long as such organisms are comprised of cells that replicate, undergo metabolic processes, and so on. Insofar as there is design in them, that design is grafted on to the knowledge we have of natural history, to take advantage of prior interactions with environments or sequences of environments. I am not referring here to the computer field of "artificial life," based on Von Neumann's other proofs, which has conceptual problems of its own.

34. Frank W. Liou, *"Rapid Prototyping Processes": Rapid Prototyping and Engineering Applications: A Toolbox for Prototype Development* (CRC Press, 2007), p. 215.

35. I wish to thank Ryan Munévar for these and other interesting suggestions and comments.

36. For example, when life in a planetary system begins to use radio and television, the alien probe may save those transmissions for a while, and then begin to broadcast them back to the originating planet, perhaps on a different frequency. This would be a way of making contact, for the inhabitants of the planet would realize that those signals were coming from outer space. In our day and age, however, such rebroadcasts may be mistaken for reruns. Nevertheless, the alien probe may "edit" *I Love Lucy* to include pictures of its home planet, or come up with another attention-grabbing stratagem. For the earliest publications on the subject, see Ronald N. Bracewell, "Communications from Superior Galactic Communities," *Nature* 186 (1960): 670–671, and "Life in the Galaxy," in *A Journey through Space and the Atom*, ed. Stuart T. Butler and Harry Messel (Nuclear Research Foundation, 1962), pp. 243–248; reprinted in *Interstellar Communication*, ed. A. C. W. Cameron (W. A. Benjamin, 1963), pp. 232–242. See also his *The Galactic Club* (W. H. Freeman, 1975).

37. R. Bracewell, *The Galactic Club* (W. H. Freeman, 1975).

38. T. Fine, "Nature of Probability Statements in Discussions of the Prevalence of Extraterrestrial Intelligence," in C. Sagan, *Communication with Extraterrestrial Intelligence* (The MIT Press, 1973), p. 360

39. Fine, "Nature of Probability Statements in Discussions."

40. Fine, "Nature of Probability Statements in Discussions."

41. This section was excerpted and published as Ch. 2 of my *Evolution and the Naked Truth*.

42. An earlier version of this section appeared in *Explorations in Knowledge* VI, no. 2 (1989). It was later reprinted as Ch. 2 of my *Evolution and the Naked Truth*, pp. 23–32. The conceptual underpinnings were first worked out in my *Radical Knowledge* (Hackett, 1981). Two other philosophers have developed views in a similar spirit: Lewis Beck, "Extraterrestrial Intelligent Life," Presidential Address, American Philosophical Association, December 1971 (Reprinted in the *APA Proceedings*, 1971, pp. 5–21); and Nicholas Rescher, "Extraterrestrial Science," Ch. 11 of his *The Limits of Science* (University of California Press, 1984), pp. 174–205.

43. See, for example, Sagan, *Communication with Extraterrestrial Intelligence*.

44. L. Margulis, *Symbiosis in Cell Evolution* (W. H. Freeman Co, 1981). Symbiosis is a plausible way for complexity to arise, but it need not be the only way.

45. See also S. J. Gould, "SETI and the Wisdom of Casey Stangel," in his book *The Flamingo's Smile* (Norton, 1985).

46. Gonzalo Munévar, *A Theory of Wonder: Evolution, Brain, and the Nature of Science* (Vernon Press, 2021).

47. New findings suggest that proto eyes are very ancient and thus that instead of convergence we have here a case of common ancestry.

48. Charles Darwin, *The Descent of Man* (Penguin, 2004).

49. P. Churchland, *Braintrust: What Neuroscience Tells Us about Morality* (Princeton University Press, 2011);P. Churchland, *Conscience: The Origins of Moral Intuition* (W.W. Norton & Co., 2019).

50. M. Bradie, *The Secret Chain: Evolution and Ethics* (State University of New York Press, 1994).

51. G. Munévar, "The Morality of Rational Ants," Ch. 11 of *Evolution and the Naked Truth*.

52. P. Singer, *The Expanding Circle: Ethics and Sociobiology* (Farrar, Strauss and Giroux, 1981).

53. D. Peterson, *The Moral Lives of Animals* (Bloomsbury Press, 2011).

54. J. S. Bernal, "The Role of Sex and Reproduction in the Evolution of Morality and Law," in *Sex, Reproduction and Darwinism*, ed. F. de Sousa and G. Munévar (Pickering and Chatto, 2012), pp. 141–153.

55. E. O. Wilson, *On Human Nature* (Harvard University Press, 1978).

56. J. Rawls, *A Theory of Justice* (Harvard University Press, 1971).

57. Mick West, "Feds' Report to Bum out UFO Believers When It Lands," *Detroit News*, May 30, 2021, p. 29A and 31A.

58. David Lamb, *The Search for Extraterrestrial Intelligence* (Routledge, 2001).

9

Space Technology and War

One of Rousseau's complaints against the arts and the sciences was that they weaken the military might of a society. "All examples teach us," he said, "that in military affairs . . . study of the sciences is much more apt to soften and enervate courage than to strengthen and animate it."[1] Perhaps it was still possible in the eighteenth century, when he wrote, to believe that the sciences were luxuries of no practical military consequences—although even then the new science had brought great advances in ballistics and other military fields. Today those who share Rousseau's suspicion of science do so for different reasons. He worried that science made us, if anything, less formidable; they worry that it makes us far too formidable.

The proponents of space exploration generally have a benign view of what the enterprise has to offer, although it is not uncommon to see space activities funded precisely because some military advantage is likely to result from them. The serious objection is not, however, that there is a connection between space technology and the military, for after all there have been times when helping the military was the right thing to do (e.g., fighting against Hitler). The objection is rather that space technology puts into human hands tools that we cannot fail to mishandle. There is an evil side to us, and so whatever discoveries science makes will eventually inflict pain and misery upon humanity. Space science and its accompanying technology are no exception. If anything, they confirm the suspicions against science in general. Is it not true that rockets have been used to kill and terrorize in the past? Is it not true that for decades they were the very means by which the entire planet could have been brought to nuclear annihilation at a moment's notice? Are they not still a great threat today?

The point is not merely that there is a connection between space exploration and grievous outcomes. The point is rather that the connection is somehow unavoidable, that you cannot have the first without the second. Thus, the more space technology progresses, the more acute the grief. This far more sweeping claim requires a look into the history of space exploration and its likely future for clues of such inherent connections with destruction

The Dimming of Starlight. Gonzalo Munévar, Oxford University Press. © Oxford University Press 2023.
DOI: 10.1093/oso/9780197689912.003.0009

and evil. Indeed, we must keep in mind that this claim often gains plausibility in the first place because of appeals to history—mainly to the role of the German V-2 rockets in World War II and of the intercontinental ballistic missiles (ICBMs) during the Cold War.

The need to examine this objection in a historical context cannot be stressed too much. Some may argue that the relevant issue is whether space technology will make war inevitable in the future, whatever its role might have been in the past. But this line of argument operates in a rhetorical vacuum. An estimate of the contribution of space technology to war should presumably be supported by reasons, by an account of the causal and probable connections involved. And how are those reasons going to be assessed? What makes causal and probable connections plausible in the first place? I submit that these difficult matters are most often influenced by the way we have learned to judge. And what has determined that learning if not our perception of how similar matters have been resolved? We appeal, that is, to our experience, and in the last analysis, to history. At the very least a brief look at history is necessary, then, to unearth assumptions which otherwise may be innocently smuggled into appraisals of the future.

The wish to explore beyond the confines of our own world is very old. By 180 CE, it received full treatment in Lukian's *Vera Historia* (True History), in which travelers go to the Moon when a giant whirlwind picks up their ship from the ocean. In 1634, the famous Johannes Kepler wrote *Sleep*, a novel about a trip to the Moon. With the advent of the Industrial Revolution, several would-be inventors tried their imagination at mechanical contraptions that could turn such trips into more than dreams, although their colorful ideas had little to do with the actual development of rocketry many decades later. In Russia there was Kibal'chich, who spent his time designing explosive devices and rocket aircraft, while ignoring his trial for blowing the czar to bits in 1881. In Germany there was Ganswindt, who worked on a hopeless steam jet to propel his spacecraft. And everywhere there were fiction writers taking their readers on trips that engineering could not yet make available.

Space Rockets

The theme of space travel was in the air, and around the turn of the century the real pioneering work was finally carried out. By then science and technology had caught up with the old dreams to the point that not one but three independent investigators provided the foundation of space rocketry. The Russian Tsiolkovsky was the earliest, and then came Goddard in the United States, and finally the most influential of them all, Oberth in Germany. The first thing these three men had to do was show that space flight was indeed possible. Their solution to this problem, as we will see, has some bearing on the issue of whether inherent connections exist between space technology and devastating war. And the problem was that, to many "experts" at the time, space flight seemed not merely far-fetched but physically impossible. Some disheartening calculations, for example, showed that however efficient the production of thrust, no rocket could raise its own mass into orbit. But as the pioneers showed, even the most sophisticated impossibility proofs could be gotten around by the very simple but ingenious idea of using multistage rockets. The first stage gives the whole rocket an initial boost and then separates. At that point a second stage takes over the task of pushing a now lighter vehicle with a now shorter distance to climb. One or two more pushes like that and the rocket achieves orbital velocity (for details, see William S. Bainbridge's *The Spaceflight Revolution*, which informs much of this chapter).[2]

A second barrier was overcome by the switch from solid to liquid fuels. Standard rockets generally used solid fuels, mostly gunpowder, which lacked both the power and the control required for space flight. Tsiolkovsky and the other pioneers soon realized, however, that several mixtures of liquid propellants, particularly hydrogen burned with oxygen, could give rockets the desired performance. This is of particular importance to our question, as we shall see. Now, a quick comparison of liquid and solid fuels shows how right Tsiolkovsky was—a most remarkable feat for a self-taught man who never gained entrance to the scientific or engineering circles of his day. Liquid propellants liberate more energy per pound than their solid competitors (the following figures were given by John Shesta of the American Rocket Society in 1936).[3] The best powder achieved 1,870 BTUs; a mixture of methyl alcohol with oxygen, 3,030 BTUs; while hydrogen and oxygen combined gave 5,760 BTUs. Even in contemporary times the differences are pronounced: the solid fuel in the typical military rocket of the 1970s produced in principle an exhaust velocity of 2,250 meters per second (m/s), a figure inferior to the 2,750 m/s that Goddard had achieved with his small rockets—decades earlier—and much below the 4,200 m/s obtained by burning hydrogen with oxygen. At least until recently, the best that could be expected of solid fuels did not measure up to the performance of liquid fuels in this respect.

In the second important respect, control, the differences are just as pronounced. The difficulty with a solid propellant is that it tends to burn until it is exhausted. You cannot just shut it off and start it again. In a liquid system, on the other hand, you may always control the amount of fluids intervening in a reaction—you may increase it, decrease it, or turn it off altogether. And you can open the valves again and thereby restart the combustion that gives you the thrust. This fine control permits the appropriate accelerations at the appropriate times to maneuver the rocket into the desired orbit.[4] And although the techniques to achieve this control were not easily acquired by rocket developers in the decades that followed, there was now a clear direction of research, as well as good reasons to think that the problems could be solved.

The three pioneers also provided basic formulas for engine performance and specified likely vehicle trajectories. Now the goals were clarified, and so were the means for attaining them. Nevertheless, the move toward space did not quicken its pace for a long time. The few who were technically competent and who took the trouble to carefully read the works of the rocket pioneers may have realized the potential involved. But most technically qualified people regarded the topic with suspicion and did not bother themselves with an investigation of it. There is nothing conspiratorial or shortsighted in that attitude. Many scientists may well be bombarded with a myriad of ideas that they have not examined in detail. Which ones should they explore? Not all of them. They cannot. And surely not those that strike them as implausible or without foundation. Time is too short for that. Science changes because not all scientists are cut from the same cloth, and thus what seems preposterous to the majority may instead strike a resonant chord in a few others. Most beginnings are therefore small, and the development of space technology was no exception.

The dreams of the pioneers were left for others to realize. Tsiolkovsky remained undiscovered and ignored. Goddard was very secretive about his own work. As Bainbridge points out, "He did not publish an account of his first 1923 engine firing until 1936, when the V-2 was already taking shape on German drawing boards. Complete reports of his experiments in the 1930s were not published until 1961, the year that Yuri Gagarin orbited the Earth."[5] This situation was apparently more the result of his peculiar temperament than anything else. According to Bainbridge, "Goddard tended to ignore the work of other men in the field, was remiss in his correspondence with colleagues, refused to share his results, and would not participate in joint projects. He seemed to want to achieve successes . . . then burst upon the world in triumph."[6] But one man alone could not achieve what took the efforts of many thousands. Such were the first quirky steps in the strange journey that took us to the Moon.

As interesting as the lives and motivations of these men were, an account that would do them justice is beyond the scope of this book. For our purposes, it is enough to say that those motivations had more to do with the liberation of the human spirit, with its excellence, than with the destruction of other human beings. In Oberth's words: "probably Mankind will even build spaceships sometime, make other planets habitable, or even establish habitable stations in space, and having become morally mature in the meantime, will bear life and harmony out into the cosmos."[7]

Oberth presented his masterwork *The Rocket into Interplanetary Space* as his doctoral dissertation in 1922 (Goddard's crucial theoretical work preceded his by ten years, and Tsiolkovsky's by twenty). He was turned down with the advice to look for a more suitable topic. He refused, and then proclaimed that he could become a greater scientist than his examiners, "even without the title of doctor."[8] His arrogance was not entirely misplaced. Within a year he had published his book, of which Arthur C. Clarke has said that it "may one day be classed among the few that have changed the history of mankind."[9] In any event, it became not only a source of inspiration but also the textbook for the German rocket experimenters. One such group, founded in 1927, the VFR—Verein fur Raumschiffahrt (Society for Space Travel)—was to prove of crucial importance.

Even though rockets had been used for military purposes since the 1200s, they had never been particularly effective as weapons. But in the 1930s, the needs of the German military began to converge with those of the VFR. The main problem for the rocket enthusiasts was the lack of financial support for their activities. They resorted to all sorts of stratagems to raise money. One member by the name of Valier sold Opel on the idea of using rockets to propel automobiles. This scheme brought in some funding until Valier was killed testing one of his contraptions.[10] The VFR leader, Rudolf Nebel, talked the city of Magdeburg into supporting rocket research that would lead to an experiment sure to make the city famous. At the time there were in Germany many crank theories supporting the view that the Earth was a hollow sphere and we lived in the inside, with our heads, not our feet, pointing to the center.

A rocket with a human observer in it could go up a few miles and decide once and for all what the real shape of the Earth was.[11] Nothing but extended survival for the VFR came of the project.

Space Rockets and the German Military

Into this opportunistic search for funds walked the German army. The Treaty of Versailles, after World War I, forbade the German development of long-range, heavy artillery. But there was no mention of rockets anywhere in the treaty. A young member of the VFR, Wernher von Braun, was able to interest the army generals in his projects, a task facilitated by the fact that he was an aristocrat and the son of the minister of agriculture. Initially, the army support was small, but as the tensions in Europe grew and the task of rearmament went into full swing, young von Braun very adeptly played the German armed services against each other (both the army and the air force saw the rocket as their domain). All the while, the story has been told, von Braun and his cohorts kept on designing the future spaceship.[12] But when it came to economic support, all the talk was of the terrifying weapon they were developing for the fatherland. Their final product—the V-2 rocket—was indeed terrifying. Many a British nightmare began with the horrible whistle of the rocket as it cut through the London night on its mission of death and destruction.

Even if the VFR used the bombing of London to finance space flight ultimately, I expect that critics will still deny them absolution. Moreover, they may argue that this episode is a clear instance of how the quest for space made for more misery than the world would otherwise have endured. And they may also point out how easy it is for scientists to sell their souls to the devil. Oberth thought that space would bring about a more harmonious future. So did the pioneers of aviation. Once in the air, human beings would realize the unimportance of the geographical and political barriers between them. In a similar vein, some believe that as humans look back at the globe of the Earth from space, they will be struck by the revelation that we are all children of the Mother Earth and therefore brothers. There lie the beginnings of real peace. Ho Chi Minh, the leader of North Vietnam during the war against the United States, sent a message of congratulations to his enemy on the occasion of the first lunar landing by the Americans. Perhaps that revelation of brotherhood will become widespread. But only folly would lead us to expect that it can guarantee peace. As for the aviators, not long after being awed by the prospects of peace, they were killing each other in dogfights and bombing cities to rubble. And despite the spectacular adventure of the Apollo flights to the Moon, the war in Vietnam continued until 1975.

V-2 Rockets—Not Efficient Weapons

Nevertheless, let us not leave the V-2 incident unexamined, for that incident gives so much plausibility to the notion that space technology is bound to bring us to grief. We assume that anything that furthered Hitler's cause was evil; therefore, the V-2 was evil. But did the V-2 further Hitler's cause? To believe Dornberger, the German general in charge of the V-2 project, it did.[13] He recalls that Hitler apologized to him for not having grasped the significance of the weapon years earlier—the first time in his life that Hitler had apologized. But the man who kept the books, Albert Speer, the minister of munitions, thought that the V-2 had been a terrible waste of manpower and resources. As he put it: "49,000 tons of explosive were dropped on Berlin alone, by which 20.9% of the dwellings were seriously damaged or totally destroyed. In order to direct the same quantity against London, we would have had to employ 66,000 great rockets."[14] Consider for a moment that Germany fired against England a grand total of 1,340 V-2s!

By contrast, one American B-17, a long-range bomber, cost six times as much as one V-2 but carried three times as many explosives and could be used many times over. It has been suggested that Germany would have been better off building airplanes instead. This is doubtful. Radar gave the allies a great advantage in defending against the German air force, and thus more airplanes may not have made the difference. Fighters could, of course, be used to defend Germany against the attacking bombers; but it is not clear that the Germans had the gasoline for those extra airplanes while, as Dornberger claimed, the V-2s used alcohol extracted from potatoes. Dornberger, however, fails to take into account the diversion of coal from the production of gasoline (the Germans made synthetic gasoline) to factories producing not only the alcohol but also the more esoteric fuels used by the rockets. Nor does he take into account that cars and trucks could and did run on alcohol. Indeed, at Peenemunde, the base for the V-2, the carburetors of trucks had been modified so they could run on rocket fuel.

Whatever the final disposition of the choice between V-2s and airplanes, there is a more straightforward comparison to determine the effectiveness of the V-2. The natural competitor of von Braun's rocket was the V-1, or buzz bomb, a pilotless plane forerunner of the cruise missile, and not a rocket designed with space in mind. According to calculations made by David Irving and later confirmed by Speer, the V-1 killed twice as many Britons for half the production cost of the V-2.[15] Toward the end of the war the British were able to shoot down most V-1s, but the Germans could have instituted easy modifications that would have preserved the success of the V-1. They did not do so because von Braun had convinced them that his big liquid-fuel rocket was the way to go. As Bainbridge concludes, "By any criteria, the V-2 was not a cost-effective weapon. It could not match the performance of much simpler weapons systems, yet drained money, materials, and talent from its sponsors."[16]

Sometimes it is suggested that the V-2 could have carried an atomic warhead (see Figure 9.1). But the Germans did not have one. And if they had, the V-2 could not have delivered it (the payload of a V-2 was less than a ton; an early atomic bomb weighed four tons). Suggestions that the Germans could have combined several V-2s for that purpose, or that they could have built a rocket capable of hitting targets in the United States do not stand up to close scrutiny. The same can be said for the notion that a von Braun design had provided an effective antiaircraft missile. But I will let the interested reader consult Bainbridge's analysis (pp. 92–122). For our present purposes the important result, at least in this case, is that space technology does not seem to have made matters worse for humankind. If anything, the case goes in the opposite direction. This, of course, may not change our moral evaluation of von Braun and his group. Bad intentions or plain callousness are in themselves worthy of blame.

After the war, the V-2 was used for scientific research at high altitude, a task for which it was very well suited. Its military use was limited, as was that of its direct descendants, the Russian T-1 and the American Redstone. And the reason was simply that a rocket was then too expensive and complicated a means of delivering conventional explosives. Nevertheless, opportunity would again come von Braun's way, now with his group forming part of the American rocket program, and for Korolyov, von Braun's Russian counterpart. At first it was the Army's interest in a rocket that might carry atomic warheads; and then in the 1950s, with the development of compact hydrogen bombs, the space rocket seemed to have become at last an effective

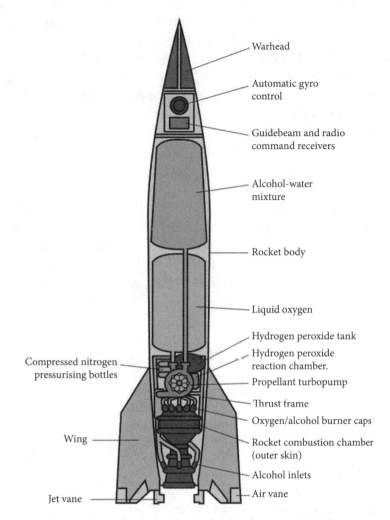

Warhead

Automatic gyro
control

Guidebeam and radio
command receivers

Alcohol-water
mixture

Rocket body

Liquid oxygen

Hydrogen peroxide tank

Hydrogen peroxide
reaction chamber.

Compressed nitrogen
pressurising bottles

Propellant turbopump

Thrust frame

Oxygen/alcohol burner caps

Rocket combustion chamber
(outer skin)

Wing

Alcohol inlets

Jet vane

Air vane

Figure 9.1. V-2 rocket schematic. (Image by Fastfission. Public domain image)

instrument of war. Thanks to the descendants of the V-2 that terrorized London, humans could now make short work of ending all life on Earth.

Space Technology and Mutually Assured Destruction

It is clear that the Cold War led both Americans and Russians to spare no effort in developing technology for war. And there is no question that the quest for space played a part in putting the human race under the threat of nuclear annihilation. But it seems to me that as in the case of the V-2, the matter is not all that simple. First, that threat would have been there independent of rockets. Bombers would have been sufficient all along. Second, for all the anxiety that threat caused, some argue that the fear of mutually assured destruction (MAD) actually prevented a third world war.[17] And third, the contribution of space technology is ambiguous.

The first point is fairly obvious. The main role of rockets was to make the time of flight so short—from a few minutes to half an hour depending on their location—that in case of a confrontation it would have been very difficult to correct any mistakes. To some critics, this is a terrible indictment of space technology. And so is the mere fact that rockets have given us one more way of making the destruction of life possible. But this is all a matter of perspective. Let us consider the two indictments separately. It is true that shortening the time from the decision to attack to the actual explosion did not allow for much flexibility. Nevertheless, the problem of inflexibility still existed with bombers and cruise missiles that could not be called back—as it was the case with the American weapons—so without rockets the world would get a reprieve of a couple of hours. Once the decision to attack was made, the result would be the same anyway. In any event, from the point of view of the supporters of MAD, the variety and efficiency that rockets offered simply made us work so much harder to make sure that we did not launch an attack. And that was on the whole for the best.

People who hold this view need only point out that the Soviet Union and the United States had such mutually repulsive political systems that a global war would have been inevitable without MAD. And even though the two superpowers constantly engaged in proxy wars, and even though many other countries insisted on adding their own ghastly contributions, during the Cold War the world endured nothing comparable to the devastation of World War II. Indeed, it is easy to imagine that a conventional war between the two superpowers, probably in the 1950s, would have dwarfed the confrontation with Hitler. Thanks to nuclear weapons, there was no world war because to start one was to commit suicide. The anxiety produced by MAD was difficult to bear. But given the alternative—and I mean a conventional, not a nuclear, World War II—there were those who would rather be anxious.

The notion is then that, with stakes so high, the world has had to be very careful. But what makes some critics despair is that an accident or a misunderstanding could have brought about the end of the world. As the means of setting the bomb off multiplied, the probability that it would happen increased. Besides, the shorter the fuse, the harder it is to evaluate whether reports that it has been lit are genuine. A country may retaliate, thinking itself under attack, in a case in which more time could have turned up a computer malfunction or a strange echo on a radar screen rather than an incoming ballistic missile. But on the other hand, the more its means of retaliation, the less a country will fear that it will not be able to get even. In such a case, a country may be more willing to take the chance that its opponent has not really launched a first strike—and more willing also to recognize an accident for what it is. The fewer its means of retaliation, the more paranoid a country may be, and thus the more likely to attack.

In this matter, as in most political matters, people hold strong opinions without hesitation. But it is seldom easy to see why the argument can cut only one way. Nevertheless, let us grant—for the sake of discussion, if nothing else—that MAD was a terrible thing, and that the world would have been better without it. Is the link between space technology and the nuclear threat therefore as firm as many believe?

That matter is not so clear either. During a short time in the 1950s, space rockets became suitable vehicles for the newly invented hydrogen bombs. But once the Russian and American space programs began in earnest, the size and power of the new rockets were generally far in excess of what the military needed. Even at the time when the first ICBMs were being built, the goal could have been accomplished without liquid-fuel rockets, that is, without space rockets. Solid-fuel rockets, it turns out, can serve the military purpose with greater reliability and safety. In fact, in several instances the space rockets were out of the question. The Navy, for example, had no use for a liquid-fuel rocket that could be fired only in near-perfect weather. The Navy feared that any rocking of the ship could make the rocket explode prematurely—on deck.

Of course, a solid-fuel rocket does not offer as much control. But then the task did not involve putting anything in orbit, and, in any event, given the devastating power of a hydrogen bomb, extraordinary precision was not at a premium. Much later when increased precision was required, new guidance systems had become available for the solid-fuel rockets. Thus, the military could have had its ICBMs without space; and to concur with Bainbridge, perfect hindsight makes us see that the military should have gone strictly with solid-fuel rockets. The direction the military actually took favored space flight and not the means of destruction. And it took that direction because space enthusiasts like von Braun could always make great propaganda out of the technological success of the V-2 and the myth that surrounded it. Without that myth, military rockets would have been very different. As Bainbridge puts it: "The military advantages of solid fuels include cost, storability, reliability, and simplicity. These advantages must be foregone by space vehicles to achieve greater power and control. By its striking superiority over contemporary solid-fuel rockets, the V-2 pointed the direct way toward space but led military technology on a detour."[18]

Another contribution of space to our actual situation is often mentioned. Satellites are an integral part in military communications and reconnaissance. But on the whole, this has been more of a benefit than not. It was precisely the existence of such satellites that made it possible for the Kennedy administration in the United States to sign a test-ban treaty with the Soviets. The days of such testing seem a remote memory today, but we must not forget that extremely powerful devices were routinely exploded in the atmosphere, with potentially disastrous effects. And even though a ban was in the interest of both parties, it was difficult to get around the suspicion that the other side would cheat. Together with seismographic methods and other techniques, reconnaissance satellites gave the needed assurances. Little can be done to hide an explosion of several megatons from a good camera overhead.

Satellites also guide "smart bombs" and give the military information needed to invade other lands. But then again, cooperation with the military need not be evil. It all depends on the enemy and the war. Besides, the main motivation for the development of "smart bombs" is to maximize the destruction of military targets while minimizing civilian casualties. This is hardly the basis for an indictment. Now, a new direct application of space technology to war may be the conversion of an airplane designed to fly in the thin Martian atmosphere into a spy plane (for it could fly high enough to avoid standard ground-to-air missiles). As we have seen, though, the spying made possible by space technology has been, on the whole, beneficial.

Let me leave behind for the moment this examination of past and present and concentrate for a while on the future. There are two main ways in which space can be seen as worsening our situation. One is that weapons more formidable may still result from space technology. The other is that by developing that technology we make the world more unstable. What those formidable weapons might be is largely unknown. A suggestion one hears from time to time is that big rocks could be aimed at the Earth from the Moon.

They would be accelerated to escape velocity by electromagnetic forces, and their course toward their Earth target would be corrected by standard guidance techniques. The energy released by the impact of a large rock could cause extraordinary damage. To mount such an attack, a country would require a rather substantial base on the Moon. This base could be underground, and presumably easy to defend. I think that Gerard O'Neill's efforts to design a mass driver to put lunar materials in orbit—basically the technology for the Moon slingshot—shows that the military applications are not so readily at hand. The rocks would have to be very big to be effective as weapons and follow a very precise trajectory to fall on the right target. Nonetheless, I suppose that sooner or later such a weapon may be feasible. But I am not sure how the situation would be radically altered. What kept ICBMs in their silos had nothing to do with how easy the silos were to defend. It had all to do with the fear of retaliation upon the society that launched them. Similar remarks should apply to the proposal to build a gigantic solar collector in the Moon to energize a very powerful beam—a death ray—which would vaporize any nation incautious enough to become our enemy.

Another possible nightmare connected with space exploration may come at the time when we make serious attempts to travel to the stars. A hydrogen bomb releases only a small fraction of the energy "frozen" in matter. To achieve the relativistic speeds necessary in star travel, we must find practical ways of releasing far larger fractions of energy. The problem is that, with such a technology, it might be possible to make bombs monstrous enough to blow our planet to bits. In that case, however, retaliation would be rendered at once impossible and redundant.

Some evidence suggests that a full-scale attack by one superpower upon the other might be enough already to destroy humanity even if there is no response. According to Carl Sagan and other researchers, a global nuclear war would radically affect the atmosphere, both by the amount of radioactive dust that would circle the planet and by making it poisonous to terrestrial life.[19] Although this possibility requires a rather pessimistic reading of "Nuclear Winter" scenarios, it should give further pause to nuclear adversaries. As our destructive power increases, to kill your enemy would be tantamount to suicide. This may well show that to build more weapons is sheer folly. We should realize, however, that there is a clear sense in which more offensive power cannot make matters worse. When you have a bazooka pointed at someone's head, bringing a far bigger bazooka does not really alter the situation all that much.

Reagan's "Star Wars" Defense

The other way in which space may presumably worsen the situation is by making it more unstable. One possibility much discussed in the 1980s was the defense system proposed by President Ronald Reagan. This system, popularly known as the "Star Wars Defense," would have employed gigantic lasers or particle-beam weapons to knock out ICBMs in flight. Normally the atmosphere would dissipate the impact of such weapons, but since ICBMs must fly in thin regions, they might be easy prey. There were several formidable problems with such a scheme. The first was that the technology required went far beyond the state of the art. The second was that several easy countermeasures were open to the other side. And the third, and most decisive, was that according to the most optimistic reliable estimates, such a defense would probably be no more than 75% effective. Since at the earliest time when the system could have been installed, each side could have owned at least 10,000 warheads, the successful 25% would be more than enough to put an end to all things human. Even 99% effectiveness would allow for incredible devastation. Although such high level of effectiveness was never in the cards, imagine that a vigorous program of research could have improved the power of lasers and the means of detection of ICBMs to the point that a 100% effective defense had been possible in a few decades. Since President Reagan's proposal included making the technology available to the other side, ICBMs would have become obsolete. In this way space would have done away with its main contribution to the anxiety of the Cold War.

President Reagan's scheme was as pointless as it was expensive: a Star Wars Defense cannot end the threat of annihilation. For the laser beams and the particles shot from low orbit could not penetrate the atmosphere to knock out also bombers and cruise missiles, or even missiles fired from nuclear submarines close to the target. That is, even with all the ICBMs neutralized, the Soviet Union and the United States had ample nuclear alternatives to destroy each other and end human civilization. In the case of a real war, our celebration of the complete success of Star Wars against ICBMs would have lasted only the few hours that it would take for a cruise missile to barely clear the last hill on its journey to us.

Whereas some hope that space technology can help us slay the dragon whose fire other technology ignited, others worry about the increasing reliance on satellites for military operations, especially now that it is possible to attack and destroy those satellites. This is seen as one more instance in which space technology brings us to the edge of disaster. But we should notice that a country whose military satellites were destroyed would feel inclined to attack, mainly because it would reasonably interpret the destruction of its satellites as the prologue to total war. A country could get away with the destruction of another's satellites only if the other could not find out about it. Since this is not so, it seems unreasonable to suppose that space technology has made matters worse.

It is true that in space we can find more reasons for fighting than we already have. If we discover a great treasure in the Moon, we may resent any attempts to take it away from us. And our satellites have become such valuable commodities that we would not like to be deprived of them. But we cannot blame space technology in this regard any more than a tribe can blame their canoes for enabling them to discover good hunting grounds downstream, grounds that may become a source of quarrel with another tribe.

Space Technology and Anxiety

By now some readers may feel that this apology of space technology is turning into the confessions of Pollyanna. If space has done much to drive technology, in some way it must have also influenced the development of the armaments that have held the world hostage to nuclear terror. However diffuse, that influence must have been there. But the most important point is this. If it had not been for technology, we would not have been in a position to destroy life on Earth. Once you achieve a certain degree of technological proficiency, total destruction becomes a real possibility. Since space will increase our technological proficiency even more, the military will have even more means of threatening the welfare of human beings. And one day something may go wrong ... Moreover, this relationship between the military and technology is inevitable because the military has the function of amassing the best arsenals that it can get its hands on. Thus, the military will always try to put technology to its own uses. Some may also fear that further advances in technology may place nuclear weapons within the reach of fanatics and terrorists. The fire that we received from the gods has been fanned by our aggression and our ambition. It may yet reduce us to ashes.

Nevertheless, this line of argument cannot be accepted on a priori grounds alone. And as we have seen, the perceptions that give it plausibility do not square clearly with an examination of the historical developments. In any event, we should be wary of endowing this presumed inevitability of the connection between science and destruction, via technology, with the full status of a law of history. In the first place, the existence of laws of history is at best a debatable philosophical thesis. In the second place, this particular "law" seems to be underwritten by some rather unclear beliefs about aggression and human nature. Whether humans are aggressive by nature still is an open question. Even if Rousseau said that men are perverse and learning only makes them worse, our understanding of human aggression is not yet at the stage where we can use it to declare laws of history.

But let me set aside these rather abstract considerations. Consider instead that not all possible technologies become reality. No one may think of some of them, for example. And even most technologies that people contemplate never are attempted. Of course, the military has a lot of money and influence. Nonetheless, that is not enough reason to conclude that our anxious predicament was inevitable. Many unfortunate coincidences were required.

Although the atomic bomb was theoretically possible, it demanded an extraordinary commitment of scientific talent and military funds. If it had not been for the threat that Hitler might be developing such a bomb, it is difficult to see why the American scientists would have been so willing to work on the project or why the Army would have thought seriously of embarking on such a quest. And the step from atomic to hydrogen bomb also required a major effort that could be justified only by the paranoia of the Cold War. But the world could have been very different. Hitler could have been killed early. Or he might have won. The Cold War could have degenerated into full confrontation, and one of the superpowers might have established hegemony over the entire world—a new grand Roman Empire. None of the technological feats in question came easy. A slightly different timing of events would have changed the political and economic environment that permitted them to be born and prosper. The use of liquid-fuel rockets as weapons is a case in point. If Oberth had listened to the advice of his teachers, his book would not have changed von Braun's life—it would not have turned him into the VFR's able envoy to the German army. Space rockets might have thus never become ICBMs.

Is it not reasonable to suppose that eventually those weapons would have been built anyway? In some historical scenarios, yes. In others, not. A person does not always buy a rifle whether he or she needs it or not, just for the hell of it. It depends on what else seems important at the time. But could space technology have been developed in a different world? And if it needed the support of the military, should we not conclude that the two must go hand in hand? Not so. Space exploration could have taken a different route, with success. For one thing, because there may be good reasons to engage in it—as we have seen in the previous chapters. And for another, because space enthusiasts may have come up with great propaganda all the same. National prestige alone, even in the absence of a cold war, can be enough of a motivation in some circumstances. As DeGaulle said in ushering France into the space age, "We must invest constantly, push relentlessly our technology and scientific research to avoid sinking into a bitter mediocrity and being colonized by the invention and capacity of other nations."[20]

Still, it is obvious that without science and technology, we would not have the capacity to destroy our planet. That I must grant. In response, however, I would like to tell a story with a relevant moral. Imagine that a group of humans is marooned in a remote island. One among them, an extremely clever scientist, figures out that a massive earthquake is going to destroy the island in one year. Scientific knowledge would prompt these people to undertake a dangerous journey that they might not survive, leading them to die sooner than if they had stayed on the island (to be successful, the trip must begin almost immediately). On the other hand, ignorance would be bliss. But only for a year. What I want to argue is that even though science may increase our chances of disaster in the near future, it may also save us from perhaps greater disasters and allow us to postpone extinction.[21] And in this task space exploration has a significant role to play. The goal of space exploration, Oberth wrote, is "To make available for life every place where life is possible. To make inhabitable all worlds as yet uninhabited, and all life purposeful."[22]

Notes

1. Jean-Jacques Rousseau, "Discourse on the Sciences and the Arts," in *The First and Second Discourses*, trans. Roger and Judith Masters (St. Martin's Press, 1964).
2. William Sims Bainbridge, *The Spaceflight Revolution: A Sociological Study* (Wiley-Interscience, 1976).
3. Bainbridge, *The Spaceflight Revolution*, p. 87.
4. Bainbridge, *The Spaceflight Revolution*, p. 90.
5. Bainbridge, *The Spaceflight Revolution*, p. 28.
6. Bainbridge, *The Spaceflight Revolution*, p. 29.
7. Bainbridge, *The Spaceflight Revolution*, p. 31.
8. Bainbridge, *The Spaceflight Revolution*, p. 30.

9. Bainbridge, *The Spaceflight Revolution*, p. 30.

10. Bainbridge, *The Spaceflight Revolution*, p. 48.

11. Bainbridge, *The Spaceflight Revolution*, p. 49.

12. Bainbridge, *The Spaceflight Revolution*, p. 116.

13. Bainbridge, *The Spaceflight Revolution*, p. 61.

14. Bainbridge, *The Spaceflight Revolution*, p. 93.

15. Bainbridge, *The Spaceflight Revolution*, p. 94.

16. Bainbridge, *The Spaceflight Revolution*, p. 96.

17. H. Kahn, *On Thermonuclear War* (Princeton University Press, 1960).

18. Bainbridge, *The Spaceflight Revolution*, p. 92.

19. R. P. Turco, O. B. Toon, T. P. Ackerman, J. B. Pollack, and C. Sagan, "Nuclear Winter: Global Consequences of Multiple Nuclear Explosions," *Science* 222, no. 4630 (1983): 1283–1292. doi:10.1126/science.222.4630.1283.

20. Quoted in *A Spacefaring People: Perspectives on Early Space Flight* (Scientific and Technical Information Branch, NASA, 1985), p. 111.

21. For an extended discussion, see Ch. 5 of my *Radical Knowledge* (Hackett, 1981).

22. Quoted in Bainbridge, *The Spaceflight Revolution*, p. 32.

10

Survival and Wisdom

H. G. Wells said once that our choice is the universe or nothing.[1] He meant
that failure to move into the cosmos would condemn us to oblivion. As I have
argued in Chapter 3 and illustrated in many other chapters, the way humans
view the world, the way we interact with the world, gives us a panorama of
problems and opportunities that will change as we strive to satisfy our curiosity (see Figure 10.1), and thus a dynamic science leads to a constantly
evolving panorama. This allows us to adapt to a changing environment or
to a variety of environments.[2] If we choose the universe, we hedge our bets
against extinction.

   If this reasoning is correct, one would expect that most reasonable people
would then find this view of science as a strong justification for the exploration and colonization of outer space. As it turns out, however, objections to
such a justification may still be presented by philosophers who question why
survival should be a value, and, in particular, why human survival should be
a value. There are also objections from some who oppose big science on ideological grounds. Thus, Wendell Berry argues that the abundance of resources
in space will produce bad character, for good character requires the discipline of finitude.[3] My purpose in this chapter is to reply to several objections
along these two lines. In doing so, I will bring up again some of the lines of
argument I presented in previous chapters, especially in Chapter 8.

Human Survival as a Value

That the survival of the human species is a value may seem beyond question to most of us, although there might be some who prefer extinction
to bad character (not that I wish to suggest here that Berry would go that
far) or to decreased chances of spiritual salvation. But even overwhelming
agreement on the value of survival might not satisfy some thinkers in their

The Dimming of Starlight. Gonzalo Munévar, Oxford University Press. © Oxford University Press 2023.
DOI: 10.1093/oso/9780197689912.003.0010

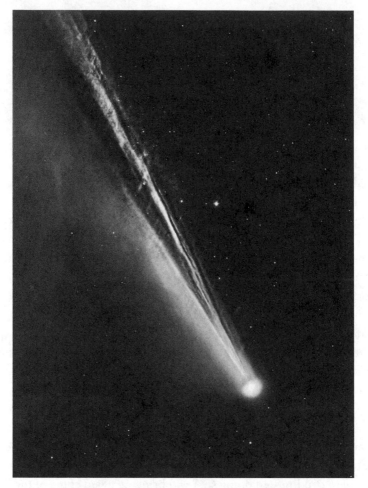

Figure 10.1. Comet C/2020 F3 Neowise. The brightest comet since comet Hale-Bopp in 1997, Neowise was visible to the naked eye in the early evening near the Big Dipper. It was only discovered in March 2020 and reached its closest approach to Earth in late June of that year. This image is about one hour of data shot with the Celestron RASA 8 (see color plate). (Image courtesy of Michael R. Shapiro)

more philosophical moments. It seems that we value survival very highly, they might say, but why should we be so keen on leaving behind imperfect creatures much like ourselves?

In such philosophical moments, questioning a value is normally taken as a demand to identify some other more basic value from which the first one is derived. This is similar to how we presumably justify actions: "This is the right action because it will bring about X and X is a good thing." But the more basic value (or good thing) that does the justifying can itself be questioned, so we then look for an even more basic value (or good thing) until eventually we arrive at a good thing that is not merely good but a self-sufficient good in itself, that is, whose goodness does not depend on anything but its own nature. We work because we get paid. Money is good because it allows us to buy food and clothes, pay the rent, and so on. We want to do those things because they contribute to our happiness. And in happiness, Aristotle thought, we find an end that is complete and self-sufficient.[4] The question "why do we want to be happy?" makes no sense. Aristotle had in mind not transient happiness, but a happy life as a whole. He also thought it was obvious that the happiness of a society was of greater value than the happiness of a single individual. And, of course, there seems to be a clear connection between human happiness and human survival.

Since this approach grounds ethical justification on a human value, human happiness, some may object that it is therefore relative to our own species. This objection seems to underpin the notion that we should not prefer the good of our own species to that of other living things in our planet, or even to the rocks of another planet. Oftentimes the objection is expressed as the view that ethics and other disciplines of value are "objective" only insofar as their laws are eternal and universal. As characterized by Peter Singer, who criticizes it, the view claims that "The laws of Ethics . . . existed before there was life on our planet and will continue to exist when the sun has ceased to warm the earth."[5] Moreover, eternal (absolute) laws of ethics seem to demand eternal (absolute) values. Thus, according to this peculiar view, a relative value such as human happiness (or human survival) cannot provide an adequate justification for our actions.

Absolute values, however, are not all they are cracked up to be. Conflict may arise between two or more absolute values. Or an absolute value may be of small significance in a particular context and thus should yield to relative values. Besides, absolute laws could in principle be derived from values that always depend on context or on subjective preference (i.e., relative). For example, consider utilitarianism (i.e., roughly, the view that the balance of good vs. bad consequences of an action—its utility—determines its rightness, given the utilities of the alternative actions). At least one version of utilitarianism would calculate utility in accordance with the values assigned by the individuals who would enjoy or suffer the consequences of the action being contemplated.[6]

I thus need not show that human survival is an absolute value, or that there must be an absolute law of ethics that gives survival a very high priority. I appeal to it in order to show that space exploration is in the interest of the species. When I point out that space exploration can save us from the dangers posed by asteroids and the Sun's becoming a red giant, I give a strong reason to pursue it.

A reason in matters of prudence, or of ethics, need not be one that appeals to an absolute ground of any kind. A reason must be a reason for action, and so it must be aimed to convince the intended audience. This is not to say that efficacy alone is sufficient to commend reasons. The fallacious reasoning of much advertisement may well appeal to the masses of the unwary but would be exposed to ridicule in less superficial disputes. In some polemics the stakes and the standards may be very high. This need not mean that some ideal is approached but that greater care must be exercised to take into account the sorts of considerations that may be brought up by all the parties concerned. And greater care must be exercised not because some of those parties are in possession of truly higher standards of reason or a more direct line to the truth—they might or might not—but precisely because we have more perspectives in play, because their diversity demands a sharper, more comprehensive case if their potential objections are to be met.

To give ethical reasons to others is then to give them reasons that take their concerns and interests into account.[7] In discussion with members of another society, as we saw in Chapter 8, we can hardly make way with claims to the effect that our customs are better than theirs because ours are ours, or because our customs appeal to us. A convincing argument would have to show them that, in some respect that they may come to see as important, our customs work better for us than theirs do for them. Or if what we really want is for them to adopt ours, we must show them that our customs will work better for them, too. If action is the intended goal of reason in matters of prudence and ethics, how can reason succeed if it cannot appeal to the audience? And what appeal can there be where the aims, desires, and interests of the audience are ignored?

In an important respect, this view preserves an element of universality, although not the peculiar ground of objectivity of so many views in ethics. As J. L. Mackie put it: "If there were objective values, then they would be entities of a very strange sort, utterly different from anything else in the universe. Correspondingly, if we were aware of them, it would have to be by some special faculty of moral perception or intuition, utterly different from our ordinary ways of knowing everything else."[8] No. The element of universality depends rather on the realization that, as Singer says, "one's own interests are one among many sets of interests, no more important than the similar interests of others."[9] In this respect, many neo-Kantian views are similar. For example, John Rawls's famous "Veil of Ignorance" requires us to put ourselves in the shoes of all those who will be affected by a decision, and to avoid results that would be completely unacceptable to us, were we in the position of those most affected (e.g., being a slave).[10]

Where the only relevant difference between my wish and yours is that it is mine, I am generally not in a position to give you reasons why you should behave as I want you to. Intelligent beings should presumably be able to detect what the relevant factors in a dispute are and discard those that are revealed as arbitrary. Or else they would go ahead with the full knowledge that their case is also arbitrary and that they have no rational claim upon the behavior of those they were trying to persuade. Practical reasoning that will not treat impartially the interests of all parties will not succeed: it cannot motivate action.

These considerations lead Singer to conclude that all rational beings should come to this process of reasoning. If so, this reasoning would have an eternal and universal aspect. For according to Singer, "Wherever there are rational, social beings, whether on earth on in some remote galaxy, we could expect their standards of conduct to tend toward impartiality, as ours have."[11] This is not to say that all rational beings would adhere to the same specific norms of conduct, for those specific norms may have developed to meet entirely divergent constraints on behavior.[12] Nor is it to say that ethical behavior among all intelligent species is possible, since such behavior requires a possible commonality of interests that may not always be there (such commonality need not be of prior interests, since in new circumstances complex intelligent beings are capable of developing new interests, surely no less than chimps and dogs can; although there is no guarantee that new, appropriate interests will in fact be developed).

In this manner, we can explain why the appeal to values is thought to provide reasons, for values themselves, as Singer points out, are inherently practical. "To value something," he says, "is to regard oneself as having a reason for promoting it. How can there be something in the universe, existing entirely independently of us and our aims, desires, and interests, which provides us with reasons for acting in certain ways?"[13] Accordingly, I point out the connections between space science and survival, intending to appeal to the interests of most normal human beings. Nevertheless, is the long-term survival of the human species really in our interest?

There are cases where survival clearly does not override other reasons (or motives) for action, and where we may agree that it should not. Cases, for example, in which someone risks her life to save her child's, or a stranger's for that matter. But all these are cases worthy of admiration precisely because we recognize that her survival was in her best interest, but that she disregarded it for the benefit of a higher purpose.

Moreover, I would venture to guess that the reason we are willing to let personal survival be overridden is that this higher purpose is somehow involved with making life better for those who remain, or even to make sure that others do remain. As this purpose expands in scope, it will ultimately cover the well-being of all humankind. And here we should not speak merely of humankind as we may find it in a slice of history, but humankind as it extends through history into the future.

Religion sometimes demands the sacrifice of lives for rather obscure goals, or for goals that only the faithful find less than revolting. And political passion is often guilty of similar motivations. But it is difficult to see how a religion or a political ideology that demanded, or permitted, the destruction of the entire human species, that would deny the future a chance, could justify itself to the most general of audiences. This is not to say that no conceivable set of circumstances could provide a reason more pertinent than the survival of humankind. Still, in the absence of a convincing account of such hypothetical circumstances, the appeal to human survival provides a compelling moral reason for choosing space.

Nevertheless, other critics may wonder about the appeal to the interests of humanity, not because appealing to interests is not enough but because they may think that "humanity" is too elusive a subject to have interests. This objection is less powerful than one may imagine. Of course, our species is not some kind of super-organism of which individual human beings are the cells. Humanity in a clear sense does not think what is best for it, nor does it recognize its interests, simply because there is no conscious subject there to think or to recognize. Individual human beings do the thinking and recognizing. That is fair enough. Moreover, the interests of human beings are individual interests. What do they have to do with the interests of humans who may live several million years hence?

Ideological Objections

Within a discussion about policy, a response should address the views of those who oppose space exploration either on social grounds (e.g., that our resources should be better used for social programs) or ideological (e.g., Berry's). It is fair to assume that those opponents will recognize that it is not only possible but also our duty to do what is best for humanity. The audience, in a figurative but still important respect, is the people of the Earth. If that were not so, what would be the point of arguing that combating poverty is more important than observing the X-ray emissions from the vicinity of possible black holes? Or of suggesting that science is not wise because in the long run it will bring us to grief? The "us" here are surely not just those of us who may hear the warning when first issued but also those in posterity whose world we may swindle by our present recklessness.

Moreover, as the social and ideological critics would likely agree, we decide for posterity to a great extent. We may plant the trees from which "our" descendants will receive nourishment and shade, or we may destroy what could have given them a fighting chance against drought and famine. It is up to them to make their own decisions, but at least the initial situation in which they will find themselves is more of our making than of theirs. Nor should we think that a society is merely an aggregate of individuals, and the species an aggregate of societies. Even if there is no super-organism, the whole does amount to more than the sum of its parts. Society is not a mere statistical distribution of individual properties. If Tony belongs to a society, he has characteristics that he could not have by himself. An advanced scientific and literate society, for example, builds libraries, universities, and laboratories, which enable Tony to educate himself for a style of life that would not exist without those institutions. The choices and opportunities open to him are not those that he could have without the benefit of the past efforts of generations that brought about the world into which he was born. No one could choose to be a modern farmer without the technology this century has provided, simply because the things a modern farmer does would not be possible otherwise. Nor could one choose to be a goalkeeper in a football team if the game did not exist. In a primitive society, it is very difficult to become a scientist, or a movie actor, or for that matter an effective critic of technology, since one will have little acquaintance with it. And in some societies dominated by religion, women do not have the right to drive a car, receive an education, or even show their faces.

What we are, what we may become obviously depends on our own efforts and talents. But it also depends on the range of choices, on the freedoms, and on the starting points that our society and culture make available to us. We do not become ourselves in a vacuum. But we also change the society by our choices, and thus we change the face of posterity, and sometimes its very substance. It is thus difficult to deny that our dialectical relationship with society imposes on us obligations of gratitude that extend to future generations. With a bit of attention, even skeptics should realize that the present generation does decide for humankind, whether unwittingly or not. And they should realize also that the choices we face today are particularly important, more so perhaps than the choices most other generations of humans had to make. And the consequences of those important choices do not affect the present generation alone. Now that space exploration has become a feasible alternative, these controversies have only been born.

Survival of the species is not a value just because it accords with evolution. In the first place, such survival is not "the goal" of evolution. Evolution has no goals. And most species that ever lived are now extinct. Survival is a value to us because without it all the other interests of the species are moot. And even though the interests of many individuals do not depend on the long-term survival of the species, the realization that certain collective actions may affect the well-being of the species makes them care how it all comes out in the wash.

According to some important contemporary views influenced by biology, it is in the nature of human beings to care about the fate of their descendants.[14] This tendency can be explained by the comparative study of life forms and their drive to ensure that their genes remain in the world even after they themselves are gone, and especially by the mechanism of kin selection and its concomitant kin altruism. Indeed, it seems that the care of the next generation has played a crucial role in the evolution of *Homo sapiens* and their morality.[15] But even if some are suspicious of such sociobiological studies and would rather speak in terms of culture, it would be difficult to deny that survival is in the interest of the species, or that our actions today may affect that interest tomorrow.

It is thus surprising that some philosophers argue that since future generations are not yet born, they do not have rights (for they are not "real"), and therefore we cannot be said to have obligations toward them.[16] Were they correct, I could not be accused of mass murder if I were to leave a large bomb hidden under the floor of the newborn wing of a hospital, timed to go off in two years, since none of my future victims would have been born at the time I hid the bomb.[17] Surely, if I know about the bomb, I have an obligation not to leave it there, even if I did not plant it myself. Likewise, we have an obligation to ensure that the corrosion of gas pipes in that same hospital wing will not reach critical mass and kill most of the newborns in ten or twenty years. Similarly, we do have obligations to ensure that CFCs no longer destroy the ozone layer so that our grandchildren will not suffer in large numbers from skin cancer. And we also have a positive obligation to put in place space systems to warn us of asteroid impacts and to deflect them, lest our descendants go the way of the dinosaurs. This is not to say that we are always looking out for the interests of the species; few of us are. But then we are practically never looking out for the interests of a stranger, although if we see him collapse on the street many of us would feel a strong impulse to come to his assistance. Similarly, the appropriate time to recognize the interests of the species is when we become aware that they are threatened. And in any event, insofar as we accept the responsibility of deciding for the species, the argument that ought to work is that which takes the interests of the species into account.

I think that the social critics should be, on the whole, very well satisfied by now. But the ideological critics may doubt the connection between space exploration and survival. I would like to consider two possible ways in which such doubts may be defended. What both alternatives have in common is the notion that the scientific approaches that would presumably save us from cosmic catastrophes in the long run will instead degrade the one habitat we know for certain is ours to the point that extinction will become far more probable.

Their first alternative approach is to face squarely the root cause of the environmental catastrophes that threaten our planet. If not prevented, those catastrophes may devastate Earth in hundreds, at most thousands of years. Neither collisions with giant asteroids nor the Sun's becoming a red giant will matter much then. And the root cause is the enormous pressure that the incredibly large human population puts on the environment, particularly when that population demands increasingly larger per capita shares of energy and other resources. Is it not obvious that if our numbers were significantly reduced, and our demands for consumption lowered, then the pressure on the environment would be relieved?

Suppose that we institute programs that would lower the total population of the planet. First, we cut it in half in fifty years. With other measures that would involve a simpler standard of living and reduced energy consumption per capita, this would certainly accomplish much to make our problems manageable. Suppose then that we continue to cut in half the population of the Earth every fifty years, until after a few centuries the impact of humans on the environment is no longer a threat to the entire planet. For some years, China took stern measures to reduce its population drastically. It seems, then, that there are alternatives to the exploration of space.

Now, how could such a momentous decision to reduce population be implemented on a global scale? The amount of coercion in China was considerable, but it seems that the strongly dominant political ideology has relented on this matter. I am not sure it is reasonable to suppose that anything short of widespread and unmitigated disaster could bring together the different hostile factions in the world to put such a program into effect globally. The disaster may come. Although by then it might be too late. And even if the struggle could still succeed, the misery visited upon humanity may be too high a price to pay, particularly if space exploration could have kept disaster at bay. It does not seem like much of an alternative after all.

The second alternative is that we better protect the Earth by nurturing respect toward it than by letting people think that we can always move onto another nest. For otherwise we imagine that spoiling our present nest is regrettable but not an insurmountable loss. Learning to live within the confines acceptable to our mother planet is a wiser policy because we would then know that we can lead dignified and fruitful lives here. By contrast, learning to live in space is only a promise. Can we bet the future of humankind on it? The greatest gift we can make to posterity is a beautiful Earth and the strength of character to live in harmony with it. In other words, we accomplish more by preserving the natural balances that have been so accommodating to human beings in the past, and by restoring such balances where modern life has already disrupted them. The result of exercising greater moral responsibility toward the world is a better world.

Space exploration, on the other hand, presumably would continue the disruption of the natural balance. If technology has already caused a mess, why should we expect better? Moreover, space exploration would be worse than a necessary evil, for it is not an enterprise that we can engage in just once before returning to a more pastoral existence. As Berry says, in condemning the scientific mind, "(1) It would commit us to a policy of technological 'progress' as a perpetual bargaining against 'adverse effects.' (2) It would make us perpetually dependent on the 'scientific' foretelling and control of such effects—something that never has worked adequately, and that there is no good reason to believe ever will work adequately." Why could it not work? Because "when you overthrow the healthful balance of the relationships within a system—biological, political, or otherwise—you start a ramifying sequence of problems . . . that is not subject to prediction, and that can be controlled only by the restoration of balance." Berry's warning is that "if we elect to live by such disruptions then we must resign ourselves to a life of desperate (and risky) solutions: the alternation of crisis and 'breakthrough' described by E. F. Schumacher."[18]

How reasonable an alternative is this to the course of action I have recommended? The first thing that deserves comment is this matter of disruption and restoration of balance. A very early and rather important disruption of natural balance took place when life was born and changed the chemistry of the planet. Another crucial and massive disruption of balance came when the oxygen liberated by life "poisoned" the atmosphere and the oceans. And this was followed by the adaptation of life to oxygen, with the subsequent destruction of the cozy arrangements between early life and the environment. Disruptions of similar magnitude were brought about by the appearance of complex organisms, by the formation of an ozone layer, which made the land available to life—would it have been better for life to stay in the oceans?—and then by the return of vertebrates to the water, which led to whales and dolphins. Ever since life began, its evolution has created new forms that have remade the environment anew, destroying the very memory of whatever balance had been struck previously, and leaving at best a few scattered fossils of what the Berrys of the time would have insisted on preserving.

The fact of the matter is that life has often created new opportunities for itself, unwittingly no doubt, and has always changed the balance between its different forms—most of which are now extinct. The biota of the planet has remade itself many times over. The natural balance of the ideological critics is merely a fiction, a temporary arrangement that would change even if there were no human beings around to mess things up. And surely life does not exhaust the range of natural causes that have brought about massive disruptions of balance. Do volcanic eruptions, droughts, and asteroids always make for small reversible changes? What may we say, incidentally, of the galactic disruption that forced the collapse of the pre-solar cloud into a planetary system? Of the earlier obliteration of what may have looked like states of cosmic equilibrium, and thus of natural balance? Which balance is it that we are morally obliged to restore?

Clearly humans are not the only creatures that transform their environment and interfere with it. As R. C. Lewontin, S. Rose, and L. J. Kamin point out:

> all living beings both destroy and create the resources of their own continued life. As plants grow, their roots alter the soil chemically and physically. The growth of white pines creates an environment that makes it impossible for a new generation of pine seedlings to grow up, so hardwoods replace them. Animals consume the available food and foul the land and water with their excreta. But some plants fix nitrogen, providing their own resources; people farm; and beavers build dams to create their own habitat.[19]

The issue is not, then, one of disrupting balance and interfering with the environment. Perfect balance can perhaps be found only right before the birth of the universe and right after its death. Even then we do not really know. And to avoid interfering with nature would be out of character for living things, while impossible to achieve anyway. The issue is rather one of interfering wisely, and of preserving (approximately) certain temporary balances that offer the best compromises for a worthwhile existence on this planet. But how are we going to achieve these goals without the kind of global and long-term knowledge that, as I have argued in previous chapters, requires the scientific exploration of space?[20] (See Figure 10.2.)

Space Exploration and the Future

I do not mean to argue that all of the environmentalists' worries are unjustified. The present rate of population growth may well be unsustainable, and the problem will have to be addressed by means I have not considered, since space exploration will not be in a position to play a significant role in the reduction of the Earth's population for a long time. If we could take a million people a year into space, which we cannot at the present, it would take thousands of years before we began to make a dent in the amount of population we have now, let alone on a population that is growing at today's dangerous rate. We have centuries, at the most. In the meantime, however, space

Figure 10.2. The Great Orion Nebula. Other than the Moon, this is the most photographed object in the night sky. The Great Orion Nebula is located about 1,400 light years away in the constellation Orion. In contrast to most nebulae, it is extremely bright, being visible to the naked eye as the middle "star" in the sword of Orion. The brightest area, known as the Trapezium, is home to intense star formation. The hot, young stars energize the surrounding medium and create the structure of the surround nebula. This image is about twenty minutes of data shot with the Celestron RASA 8 (see color plate). (Image courtesy of Michael R. Shapiro)

science may help us monitor the pressure on the environment and avail our planet of some of the resources of the solar system, such as electricity from solar power satellites. Thus, my point is not that environmental concerns are unimportant. Just the opposite. They are very important, and we should take the steps necessary to make decisions based on the most comprehensive picture of the Earth's environment we can obtain. And to do so, we need to understand what a planet is by engaging in comparative planetology, and then to observe the Earth from space in the context of that understanding.[21] Only then we can pay proper and fruitful attention to the interests of our species. What I argue against is the rigid demand to act on myths about nature that have little more than environmentalist mysticism in their favor.

Moreover, let us remember that in the long run the workings of nature, if we do nothing about them, are bound to create first a most unpleasant world for our descendants and then bring extinction upon them. Having science is no guarantee that those disasters will not happen. We cannot be assured that the desired level of knowledge is possible within whatever time limits infringe on our future. Nor can we be certain that just because we have that knowledge we will choose according to the best interests of the species. But we can be sure that without the global knowledge that requires space science, we will simply have no choice to make. Our descendants will suffer for it, and eventually our species will disappear at the earliest cosmic inconvenience.

A species full of self-hatred may well choose such a path. But is the appeal of presumed atonement the most fitting choice for us to make? That ascetic choice may buy us a bit of time, but for what? In the long run, it leads us straight into a mass grave. The other choice, the one that really lets nature run its course, offers the opportunities for expansion and diversification that so far life has been fortunate to procure for itself. It is the choice that really lets nature run its course because it recognizes that we, too, are part of nature. There is no question in my mind that in this case the way of nature has the potential for greater wisdom.

Sometimes it is said that a little bit of knowledge is dangerous, and that since we are not likely to have complete knowledge through space, or any other kind of exploration, we are better off not embarking on this scientific enterprise in the first place. But we have seen clearly, I hope, that even though a bit of knowledge can be dangerous, there is no long future in ignorance.

The decision is, of course, not mine to make. My intent has been to bring before the reader considerations relevant to these large issues. I do hope that, as H. G. Wells said, "Life, forever dying to be born afresh, forever young and eager, will presently stand upon this earth as upon a footstool, and stretch out its realm amidst the stars."[22]

Notes

1. Taken from Carl Sagan in http://www.imdb.com/title/tt0081846/quotes? qt¼qt0317500

2. Previous published as Gonzalo Munévar, "Space Exploration and Human Survival," *Space Policy* 30 (2014): 197–201. For work leading to these views, see also Gonzalo Munévar, "Philosophy, Space Science and the Justification of Space Exploration," *Essays on Creativity and Science*, ed. Diana M. DeLuca (HCTE, Hawaii, 1986), pp. 89–96; "Space Colonies and the Philosophy of Space Exploration," *Space Colonization: Technology and the Liberal Arts*, ed. C. H. Holbrow, A. M. Russell, and G. F. Sutton (American Institute of Physics, Conference Proceedings 148, 1986), pp. 2–12; "A Philosopher Looks at Space Exploration," as Ch. 13, Gonzalo Munévar, *Evolution and the Naked Truth* (Ashgate, 1998), pp. 169–179; "Humankind in Outer Space," *The International Journal of Technology, Knowledge and Society* 4, no. 5 (2008): 17–25.

3. See Berry's contributions to *Space Colonies*, ed. Stuart Brand (Penguin Books, 1977), pp. 36–37 and 82–85.

4. Aristotle, *Nicomachean Ethics*, 2nd ed., trans. Terence Irwin, Bk. 1, Ch. 7 (Hackett, 2000).

5. Peter Singer, *The Expanding Circle: Ethics and Sociobiology* (Farrar, Strauss and Giroux, 1981), p. 105.

6. Many utilitarians, however, assign to pain and pleasure absolute values, positive or negative respectively.

7. In this I follow Singer in his *The Expanding Circle*.

8. Quoted in Singer, *The Expanding Circle*, p. 107.

9. In this I follow Singer in his *The Expanding Circle*, p. 106.

10. John Rawls, *A Theory of Justice* (Harvard University Press, 1971).

11. Rawls, *A Theory of Justice*.

12. This point would apply also to the development of science: We and aliens, if they exist, are likely to have different sciences. See my "Human and Extra- terrestrial Science," *Explorations in Knowledge* 6, no. 2 (1989): 1–9, reprinted as Ch. 2 of *Evolution and the Naked Truth*.

13. Singer, *The Expanding Circle*, p. 107.

14. Gonzalo Munévar, "The Morality of Rational Ants," in *Evolution and the Naked Truth* (Ashgate, 1998), pp. 131–147.

15. Julia Sandra Bernal, "The Role of Sex and Reproduction in the Evolution of Morality and Law," in *Sex, Reproduction and Darwinism*, ed. F. de Sousa and G. Munévar (Pickering and Chatto, 2012), pp. 141–152.

16. Robert M. Adams, "Existence, Self-Interest, and the Problem of Evil," *Nous* 13 (1979): 57; Derek Parfit, "On Doing the Best for Our Children," in *Ethics and Population*, ed. Michael Bayles (Schenkman, 1976), pp. 100–102. Thomas Schwartz, "Obligations to Posterity," in *Obligations to Future Generations*, ed. Richard Sikora and Brian Barry (Temple University Press, 1978). For a discussion, see Robert Elliot, "The Rights of Future People," *Journal of Applied Philosophy* 6, no. 2 (1989): 159–169.

17. Presuming that I time it so precisely that no nurses or visitors will be killed.
18. Berry, *Space Colonies*, p. 83.
19. R. C. Lewontin, S. Rose, and L. J. Kamin, *Not in Our Genes: Biology, Ideology, and Human Nature* (Pantheon Books, 1984).
20. Munévar, "Humankind in Outer Space."
21. Gonzalo Munévar, "Why Should Philosophy Influence Science Policy: The Case of Space Exploration," *Explorations in Knowledge* 13, no. 1 (1996): 9–17.
22. H. G. Wells, *The Outline of History* (1920). www.spacequotes.com/.

Index

For the benefit of digital users, indexed terms that span two pages (e.g., 52–53) may, on occasion, appear on only one of those pages.